LOOKING
FAR
NORTH

Also by William H. Goetzmann

Army Exploration in the American West,
1803–1863

Exploration and Empire:
The Explorer and the Scientist in the
Winning of the American West

The Mountain Man

When the Eagle Screamed:
The Romantic Horizon in American Diplomacy,
1800–1863

The West as Romantic Horizon
(*with Joseph Porter*)

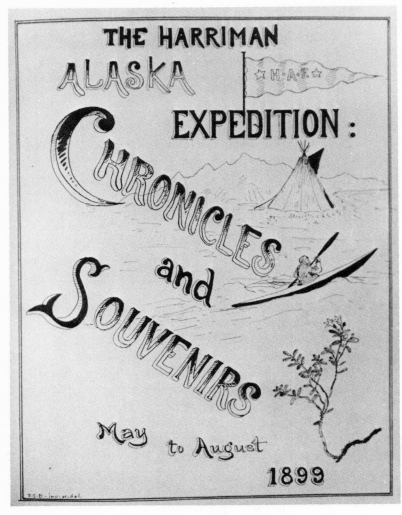

THE HARRIMAN
ALASKA
☆ H·A·E ☆
EXPEDITION:
CHRONICLES
and
SOUVENIRS
May to August
1899

Cover to E. H. Harriman's Personal *Souvenir Album*.

LOOKING FAR NORTH

THE HARRIMAN EXPEDITION TO ALASKA 1899

*William H. Goetzmann
and Kay Sloan*

Princeton University Press
Princeton, New Jersey

Published by Princeton University Press,
41 William Street, Princeton, New Jersey 08540

In the United Kingdom: Princeton University Press,
Guildford, Surrey

First Princeton Paperback printing, 1983

LCC 82-61034
ISBN 0-691-00591-5 pbk.

Photographs on pages 7, 122, 169, and 203 courtesy of the
Averell Harriman Collection; all other photographs
courtesy of The Bancroft Library

Printed in the United States of America
Set in V.I.P. Janson
Text designed by Kathryn Parise
Reprinted by arrangement with The Viking Press, Inc.

To
LaVerne and Andrew G. Sloan
and to
the faculty and students, past and present,
of the American Civilization Program at the
University of Texas, from whom we have learned so much

John Muir to: "Girls four, Mary, Cornelia, Elizabeth and Dorothea—'The Big Four' who with Carol and Roland Harriman, the 'Little Two,' kept us all young in the never-to-be-forgotten H.A.E."

"Nearly all my life I wandered and studied alone. On the *Elder*, I found not only the fields I liked best to study, but a hotel, a club, and a home, together with a floating University in which I enjoyed the instruction and companionship of a lot of the best fellows imaginable, culled and arranged like a well balanced bouquet, or like a band of glaciers flowing smoothly together, each in its own channel, or perhaps at times like a lot of round boulders merrily swirling and chafing against each other in a glacier pothole.

"Just to think of them!! Ridgway with wonderful bird eyes, all the birds of America in them; Funny Fisher ever flashing out wit; Perpendicular E. erect and majestic as a Tlingit totem pole; Old-sea-beach G. hunting upheavals, downheavals, sideheavals and hanging valleys; the artists reveling in color beauty like bees in flower beds; Ama-a-herst tripping along shore like a sprightly sandpiper, picking kelp-bearded boulders for a meal of fossil molluscs; Genius Kincaid among his beetles and butterflies and 'red-tailed bumble bees that sting awful hard'; Inuit Dall smoking and musing; flowery Trelease and Coville; and Seaweed

Saunders; our grand big game Doctor, and how many more! Blessed Brewer of a 1000 speeches and stories and merry ha-has, genial John B. who growled and scowled at good Bering Sea and me, but never at thee. I feel pretty sure that he is now all right at his beloved Slabsides and I have a good mind to tell his whole Bering story in his own sort of good-natured, gnarly, snarly, jingle, jangle rhyme."

"Kill as few of your fellow beings as possible and pursue some branch of natural history at least far enough to see Nature's harmony."

August 30, 1899

"Hardly less important than the actual fruit of the expedition is its value as a sign-post to our multi-millionaires. A little while ago a Western man of vast wealth was heard to complain to a friend that he did not know how to spend his money satisfactorily. We venture to believe that this is an embarrassment commoner than people often realize. Only the distorted imagination of the professional agitator and the sensational press really conceive of our wealthy men as a class apart . . . With the great increase in the number of people possessing large fortunes, there has come even a greater growth of the sense of responsibility. Mr. Harriman's Alaska Expedition and its magnificent results seem to indicate one true solution of this problem, and it is to be hoped that the great field lying open before them will prove attractive to our American rich men who are tired of the game of mere money-getting."

—William Healey Dall
"Discoveries in Our Artic Region,"
The World's Work, December 1900

PREFACE

Though this is the story of an exploring expedition—the Edward H. Harriman Expedition to Alaska in 1899—it is also an inquiry into the ironies of perception. The railroad magnate Harriman's voyage to Alaska was perhaps the last grand expedition of the nineteenth century. In its attempt at comprehensiveness, it is characteristic of the two centuries of scientific exploration that had gone before it, which form a very distinctive "second age of discovery," quite different from that of the time of Christopher Columbus. Harriman took with him a team of the nation's leading scientists that dwarfed anything that, for example, Captain Cook, even with the help of the Royal Society, could muster slightly over a century earlier. He also took with him two well-known artists and two photographers, one of whom, Edward S. Curtis, would soon become one of the giants of his profession.

Socially, Harriman's floating "think tank" of scientific explorers formed an elite, representative of America's version of "the genteel tradition," and many of the "high jinks" aboard his luxurious vessel, the *George W. Elder*, then thought to be hilarious, today seem merely quaint, as do the high patriotic pronouncements concerning the United States and the recent war with Spain. It was the end of a different age, with different mores, though it would be too simple to term it, as some have, an "Age of Innocence." Rather it was an age of strange binocular vision that was at times replete with irony. Essentially the Harriman expeditioners saw two Alaskas—one, the stunning, pristine land of forests and mountains and magnificent glaciers, the other, a last frontier, being invaded by greedy, rapacious, and sometimes pathetic men, often living out a false dream of success. This vision of the "two Alaskas" was in many ways a reprise of the frontier experience in the then lower forty-six states, where the land was exploited as rapidly as possible and the Indian was dispossessed and trampled under the juggernaut of "civilized" progress best exemplified by phase capitalism. Literally hundreds of thoughtful observers, from Fenimore Cooper and George Catlin onward, had lamented the process, though in 1893 the historian Frederick Jackson Turner termed it "our national epic."

Harriman's coterie of scientific experts, artists, photographers, literary men, and conservationists all noted the existence of "the two Alaskas," but few of them expressed the strong indignation so characteristic of the muckraking journalists who would follow them. Instead, they often took refuge in "high-mindedness"—offering stern lectures on the lack of character in the men who sought gold, and then turning to the contemplation of the birds and the flowers and the grand mountainous vistas. Even though the facts of Alaska's history, presented in a shipboard lecture by William Healey Dall, clearly indicated that in human times Alaska had never been anything but the prostrate victim of man's pillage, neither Dall nor his audience did anything but take the news in stride. For the most part, intent upon their individual investigations, and in general approval of the ethics of the frontier phase of capitalism, they characteristically looked the other way or saw some good in the wonders of human ingenuity best symbolized by the massive Treadwell Mine and the immense, odoriferous salmon canneries. Now and then John Muir grumbled, John Burroughs clucked, and George Bird Grinnell worried about what exploitation was doing to the native Americans, but on the whole they lived in a transcendental dreamworld. They looked right past exploited Alaska to the pristine wonders of nature. They lost themselves in the surreal experience of mile-long frozen glaciers and sparkling fjords, where immense forests ran down to an ice-dotted sea encircled by the enchantments of glacial ice palaces. They noted volcanic Bogosloff Island, the wildlife abundance of the lonely Pribilofs, and they landed in Siberia as if on an unknown continent on the shores of an unknown sea. At night aboard ship with an air of unreality they "danced by the light of the moon"—or the long unreal Arctic daylight.

In short, they succumbed to the spell of Alaska just as completely as, though in far different fashion from, the proletarian gold seekers whom they scorned or pitied. Nothing underscores their bizarre, surreal vision of Alaska better than Edward S. Curtis' photographs. With the exception of obligatory souvenir shots of squalid but quaint Indian dwellings and posturing expedition members, he concentrated on capturing the vast wonders of the Alaskan landscape in a way that prefigures the work of Ansel Adams. He never recorded the sordid details of the gold rush and, like most of the expedition members, he looked right past the two Alaskas in his photographic reveries. And yet, viewing his photographs and reading the accounts of the expedition, we know that the two Alaskas were there before them. It was a question of interest—and of perception.

Today, of course, the problem of the "two Alaskas" is with us in a particularly acute fashion. Resource bonanzas, principally oil strikes, have taken the place of the gold rush. Pipelines, not ambitious railroad schemes, are the issue. Politically, local sovereignty and dreams of local prosperity compete with what a sizable number of citizens, more alert to environmental problems, deem the national interest. Native Americans, as well as environmentalists, are better organized now, in part thanks to the later pioneering efforts of three members of the Harriman Expedition—Edward S. Curtis and George Bird Grinnell, who stood staunchly for the Indian as well as for conservation, and John Muir, who founded the Sierra Club. And while Congress debates the thorny problems of the "two

Alaskas," attempting to promote both economic progress and wilderness preservation at the same time, numerous present-day writers extol the value of maintaining the wilderness forever, unspoiled. The same themes, the same contradictions, run through the diaries, letters, and recorded conversations of the men of the Harriman Expedition of 1899. This is one of its great, almost inadvertent, perceptions, valuable even today as a point of comparison. History does not always offer consolation.

Superficially the cruise of the *George W. Elder* was comparatively uneventful. The ship did not strike an iceberg. No explorer was marooned in the frozen Arctic. No international incident was provoked. Likewise, the men of the expedition did not discover any unknown lands beyond the Harriman Fjord, with its numerous hidden glaciers. But a closer look at the day-by-day activities of the expeditioners reveals quite clearly the ambiguities inherent in the Victorian American mind as the nineteenth century drew to a close. The dual vision of Alaska as a wilderness to be preserved and a frontier to be exploited suggests the main thread of ambiguity that ran through the minds of the expedition's members. Most of them were high-minded, accomplished Victorian individuals, to whom the voyage was an opportunity for service to science and to the country. They shared a common patriotism that was occasionally not uncritical. They also shared a love of nature, but they viewed it through a number of different lenses. Each had his particular field of expertise; each had his enthusiasms. In part, this expertise, this growing specialization of perspective itself, governed one's position on the "two Alaskas" question. Bernhard E. Fernow, a forester trained in Germany, where scientific tree-farming was the vogue, quite naturally examined the possibilities for lumbering in the new territories. W. B. Devereux, the mining engineer, perforce was concerned with Alaska's mining future and with the technology of mining on a unique frontier. Dr. A. K. Fisher was downright belligerent in his search for birds, while Harriman, beyond looking at the vast country for economic opportunities, seemed primarily concerned with bagging a bear—preferably a Kodiak bear because it was the largest of the species. Charles Keeler, the California poet, was largely at a loss as to what *he* was supposed to do, so like Muir, Burroughs, and Louis Agassiz Fuertes, he simply enjoyed the birds and flowers and the spectacular mountain scenery. G. K. Gilbert and Henry Gannett were nothing if not systematic in the pursuit of their specialties. Grinnell had the plight of the Indian on his mind—a problem he must have shared in many a conversation with Curtis, who promised to send him all the Indian pictures that he took on the voyage. Thus, professional interests and personal inclination governed what each man brought to the voyage and the way in which each saw Alaska.

Throughout the voyage, too, Harriman and the designated members of his committees attempted to maintain a civilized, cultural tone that was not lacking in conviviality. In this they succeeded remarkably well. The *Elder* was, as they put it repeatedly, "a floating university," vaguely resembling the Washington Academy of Sciences, and the expeditioners took a genial interest in the lectures delivered by their colleagues. Experienced Alaska hands like William Healey Dall held them enthralled, and the Harriman children could certainly say, paralleling Herman Melville, "The *George W. Elder* was my Yale College and my Harvard."

A closer examination of the personal diaries and letters of the expedition members also reveals their human qualities in a way that sometimes provides insight, not only into their personalities, but also into the emotional makeup of the typical eminent Victorian. In some ways, the off-hand comments or deeds set down in diaries not meant for publication are like snapshots into the psyches of the expedition members. They are not always flattering to the writer, nor to the object of his comment. They often express hidden tensions, personality clashes, and significant philosophical differences such as those that are revealed, as any psychiatrist knows, in the casual re-

lation of seeming trivia. Not all the public statements of the
expedition members were genuine expressions of their feelings
and insights. Moreover, in his general narrative, the kindly
John Burroughs quietly excised what he regarded as observations too crude or frank for Victorian eyes.

Still, in all, a good spirit of camaraderie existed aboard the
George W. Elder. It was a kind of Camelot afloat on Alaskan
seas, where the savants were enabled to go on their individual
or collective quests, however strange they might have seemed.
In short, the *George W. Elder* was "a capital ship for an ocean
trip," as the reader will undoubtedly see.

We hope at least one other form of perception emerges from
our story—the view of Alaska through the eyes of emergent
modern science. Gone, with the exception of John Muir and
John Burroughs, was the all-purpose naturalist. Instead, science had become highly specialized, as the thirteen published
volumes of the Harriman Expedition Reports attest. To the
layman and the entrepreneur, science was still thought of as
something practical—capable of being applied. In most minds
the geographer Henry Gannett, the mineralogist Benjamin K.
Emerson, and the mining engineer W. B. Devereux were the
most valuable members of the expedition. They provided the
clues and the "know-how" to reach exploitable resources. E. H.
Harriman, of course, was admired as a man "who got things
done." But most of the scientists pursued their own speciali-
ties, even to the extent of collecting nine-foot worms, a particular enthusiasm of Wesley R. Coe. William Brewer stuck to
meteorology, A. K. Fisher and Louis Fuertes to birds; the great
geologist G. K. Gilbert hewed to close examination of the glaciers. But the parts did not add up to the satisfying geographic
whole that the nineteenth century had come to expect. Indeed
Gilbert, staunch empiricist that he was, ended up reporting in
abstractions and theories far beyond the common-sense scientific tradition. Science and modes of scientific perception were
clearly changing. The problems were more sophisticated and,
on the part of some, there was the frank recognition that the
process of observation altered the data. This was best stated

by Gilbert, who said simply and without a trace of irony, "an observer . . . sees what he has eyes to see."

Despite all its blindness and scientific specialization the Harriman Expedition did produce important scientific results, which we have assessed in a concluding chapter. Like most such expeditions Harriman's venture did not really end with the return to Seattle in the late summer of 1899. For the next twelve years C. Hart Merriam oversaw the publishing of a long series of official reports, which even today form a scientific reference point for Pacific Coast natural science. And for some, like Edward S. Curtis, who was persuaded aboard ship by George Bird Grinnell, it marked the start of a lifelong work—in this case on behalf of the American Indian. For this reason we have chosen to provide a conclusion that samples the Harriman Expedition's legacy to posterity in its most significant forms—science and photography.

Most of all, however, we hope that today's reader will come to know and understand these Victorian people as we did, and that he or she will experience Alaska not only as they did, but with an eye to present concerns as well. To us the Harriman Expedition is more than history. It is *déjà vu*.

ACKNOWLEDGMENTS

Because Edward H. Harriman's papers were destroyed in a warehouse fire in 1913, our search for materials on his expedition has been extensive and has taken on something of the excitement of a detective story. As a consequence we have incurred many debts and been the recipients of many kindnesses, which we wish to acknowledge here.

Though it had been a lingering interest for many years, the Harriman Expedition took on new importance for us when, upon our recommendation, the University of Texas Humanities Research Center acquired a copy of the rare Harriman *Souvenir Album*. We instantly recognized that we had access not only to an almost forgotten episode in the history of American exploration but also to a rich cache of Edward S. Curtis photographs that were quite unlike his more famous work. Thus the foresight of Warren Roberts, then Director of the Humanities Research Center, in purchasing the *Souvenir Album* generated both the problem and the opportunity that made our project interesting and, we hope, meaningful.

The staff of the immensely rich photographic collection of the University of Texas, principally Roy Flukinger and Trudy Prescott, have been enormously helpful to us. For actual reproduction of the photographs in this book, we must thank the Bancroft Library of the University of California, Berkeley, a most enlightened institution. We also wish to thank the Smithsonian Institution, and in particular William Deiss, Josephine Jameson, and Alan Bain, as well as Alan Levitan of the California Academy of Sciences, who made our working copies of the expedition photographs. Michele L. Aldrich of the American Association for the Advancement of Science was an invaluable aid in this process, in addition to helping us locate Harriman Expedition manuscripts in Washington and, for that matter, all across the country. She ranks as one of this or any other generation's outstanding researchers, and over the years she has set a new standard for scholarly friendship.

Other members of the Washington, D.C., scholarly community who have helped us include Paul T. Heffron and Jerry L. Kearns of the Library of Congress; Paul H. Oehser, historian of the Cosmos Club; and Michelle J. Epstein of the National Audubon Society.

On the other side of the continent, Robert Monroe, Dennis Christensen, and Karyl Winn of the University of Washington Library were most helpful, both in providing us with materials and guiding us through the intricacies of Northwest Coast photographic history. In this respect Frank L. Green and his staff, especially Jeanne Engermann, at the Washington State Historical Society in Tacoma were extremely helpful, as was the history department at the Seattle Public Library. We also wish to acknowledge the kindness of Mr. Asahel Curtis, Jr., of Seattle and Mrs. Florence Graybill, Edward S. Curtis' nephew and daughter, respectively. We thoroughly enjoyed the time spent with these fine people, and we appreciate the information communicated to us by Manford E. Magnuson, Edward S. Curtis' son-in-law, as well as the knowledge shared by Victor Boesen and Mick Gidley, whose previous works on Curtis are well known. A perfectly splendid day spent with Mrs. Lois Flury at picturesque Port Townsend, Washington, at the tip of Curtis Country—the Olympic Peninsula—provided a whole new dimension to our knowledge of the famous photographer.

The Frederick Webb Hodge and George Bird Grinnell papers at the Southwest Museum in Los Angeles were most crucial to our project, and we thank Mrs. Ruth Christensen for her help in making them available to us. In like fashion we thank Hilda Bohem of the U.C.L.A. Research Library, the staff of the Bancroft Library at Berkeley, and the staff of the Huntington Library. One of us wishes particularly to thank the late Ray Allen Billington of the Huntington and Martin Ridge of that exquisite library for granting a fellowship that, though not taken up, nonetheless drew him to the Huntington, where the John Burroughs notebooks proved critical to this study.

Nancy Fappiano of the Yale University Library also provided invaluable aid, as did William Emerson, Director of the Franklin D. Roosevelt Library at Hyde Park. Though he is a "presidential scholar," Dr. Emerson's knowledge of things cultural proved to be as wide as his generosity in steering us to important collections. Mr. Thomas Lange at the Morgan Library also gave us valuable clues as to Curtis materials, and we profited greatly from discussions at the Buffalo Bill Summer Institute in Western Studies at Cody, Wyoming, with Mr. Alvin Josephy of *American Heritage* and the Museum of the American Indian. Professor Ronald Limbaugh, Kathryn Kemp, and Bernice Lamson at the Holt-Atherton Pacific Center for Western Studies, University of the Pacific, were most helpful in guiding us through the Muir Papers. We are grateful to them and to the John Muir Estate.

Perhaps our most interesting experience was interviewing the only surviving member of the Harriman Expedition, Governor Averell Harriman of New York. Governor and Mrs. Harriman, together with their assistants Margaret Chapman and Gloria di Pietra, were most gracious in locating and making available to us Governor Harriman's collection of mementos from that voyage he made so long ago.

Financial support for the project came in part from the University of Texas Research Institute, the National Endowment for the Humanities, and the Center for Advanced Study in the Behavioral Sciences.

The chore of typing drafts of a manuscript written by two authors separated by eighteen hundred miles was a formidable one. Our thanks go to Corky Rosello, Meredith Racicot, Veronica Taylor of the University of Texas American Studies Program, and Deanna Dejan and Heather Maclean of the Center for Advanced Study in the Behavioral Sciences, who seem to have virtually mastered the intricacies of the word processor. Margaret Amara, the Center's librarian, and her assistants Pat Knobloch and Bruce Harley couldn't have been kinder or more helpful.

Penultimately we wish to tender our most sincere thanks to those who rendered a vast range of services to us, such as providing critical bits of advice and information, making a reconnaissance of the Harriman Collection, instructing us in photography, putting us up, or just plain putting up with us. These splendid friends include Shannon Davies, Gail Caldwell, Jeff Meikle, Louis Black, E. Donald Hirsch, Linda Farrell, Jay Martin, Raymund Paredes, Ricardo Romo, and Will N. Goetzmann.

And finally one of us, at least, would like to offer thanks to Mewes Goetzmann, who played the patient, wise Penelope in our odyssey with Harriman, Curtis, and company aboard the *George W. Elder* bound for Alaska and the Bering Sea.

Stanford, California   W.H.G.
Austin, Texas   K.S.

CONTENTS

Preface
xi

Acknowledgments
xix

THE JOURNEY FAR NORTH
I

Chapter One
The Dynamics of Philanthropy
3

Chapter Two
Across the Continent in
Pullman's Palace Cars
16

Chapter Three
"The Best Object Lessons
to Be Found on the Coast"
31

Chapter Four
The Stuff of Legends:
Skagway, the Gold Rush, and
Dead Horse Trail
55

Contents

Chapter Five
John Muir's Country
68

Chapter Six
"Cradled in Custom"
86

Chapter Seven
Malaspina's Mistake:
Yakutat Bay
95

Chapter Eight
"The Map's Void Spaces"
105

Chapter Nine
"Dog Dirty and Loaded for Bear"
116

Chapter Ten
The Siberian Connection
129

Chapter Eleven
"I Don't Give a Damn
If I Never See Any More Scenery"
146

Chapter Twelve
"The Taking of the Totems"
161

Chapter Thirteen
"A Lapse of Time and a
Word of Explanation"
171

Contents

EPILOGUE:
MR. HARRIMAN'S LEGACY
179

Chapter Fourteen
Edward S. Curtis' Alaskan Vision
181

Chapter Fifteen
Scientific Results
193

Appendix:
Members of the Harriman Alaska Expedition
207

A Note on the Sources
213

Notes
220

Index
237

THE JOURNEY
FAR NORTH

CHAPTER ONE

THE DYNAMICS
OF PHILANTHROPY

Huddled in rain slickers and under umbrellas, a hardy crowd of spectators jammed the old Seattle wharf on May 31, 1899. Through a light fog and a steady, drizzling rain, the onlookers jostled one another, stood on tiptoe, strained to catch even a glimpse of the strangest group of "celebrities" yet seen in the far Northwest. They had come to see, not P. T. Barnum, Jenny Lind, General Tom Thumb, or even the familiar party of eager Klondike gold-seekers, but a group of scientists and artists bound for America's last fronter—Alaska. Most people in the crowd could scarcely identify the individual members of the purposeful group, though they represented the elite of the country's scientific establishment, several outspoken nature writers, and one veteran of Major John Wesley Powell's famous voyage down the canyons of the Colorado River. The citizens of Seattle knew something important was happening, however, because there on the dock stood Edward H. Harriman, the great railroad tycoon, coolly directing the loading operations.

Such an eminently successful man did not waste time—or money. Yet there he was, intent upon loading a mountain of supplies aboard a luxurious steamship, the *George W. Elder*, chartered specifically for his expedition to the north. As he moved rapidly about the motley assortment of people and supplies, issuing orders in crisp, efficient fashion, people could not help wondering just what purpose lay behind this lavish expedition. Rumors spread everywhere, the most prominent being that the intrepid Harriman intended to build a railroad around the world, and he was heading for Alaska to see if the project required a tunnel under the Bering Strait or merely a fifty-mile-long suspension bridge like the kind the Roeblings had hung between Brooklyn and Manhattan. For America's industrial giants nothing was impossible.

And so the citizens of Seattle stood in the rain for hours staring in fascination as luggage, camera equipment, an organ, a piano, hunters' traps, tents, canoes, rolls of painters' canvas, surveyors' apparatus, boxes of guns and ammunition, cases of champagne, lantern slide projectors, and all manner of paraphernalia were carried up the gangplank while cows, turkeys, chickens, and horses filed into the hold in a scene reminiscent of Noah's Ark. As if to confirm the impression, after the supplies were loaded, the scientists and artists who were to make up the expedition filed up the gangplank and into the ship. There were two men in each discipline—two geologists, two botanists, two zoologists, two foresters, two mining engineers, two photographers, two artists, and even two famous preservationists, the celebrated John Burroughs and the "old man of the mountains," that veteran Alaskan explorer, John Muir, who stood out because his scraggly, unkempt, casual appearance contrasted with the smart outfits of many of the other explorer-scientists. Some waved heartily to friends and relatives as they boarded the ship. Cheers rose from the crowd, especially so when Mrs. E. H. Harriman and five Harriman children, the three youngest clad in nautical apparel, arrived by carriage at the gangplank and boarded the ship. Like Noah of old, E. H. was taking his family on what he disingenuously called a "vacation." Nobody, of course, believed that. Who would vacation with a gaggle of "scientifics" and at such expense? Something important was taking place. But Harriman, as was the custom of tycoons, kept his own counsel, controlled all interviews with the press, and would only say that he was taking a "vacation" in the interests of public philanthropy, which he hoped would add to the ever-mounting total of human knowledge in "our enlightened age."

Edward Harriman's career stood at a turning point. A rapid ascension to power in the competitive railroad world had recently brought him, at fifty years of age, enormous wealth and public recognition. After years of steadily developing the Illinois Central Railroad as its financial manager, Harriman's career suddenly blossomed in 1897 when he made a bold move to control the Union Pacific Railroad. Harriman's covert political manipulations blocked the progress of the Union Pacific until the railroad's directors offered him a seat on the board of directors in 1897. His mastery of the coercive tactics wielded by the rulers of the nation's railroads had paid off handsomely; only a year later, in May of 1898, the industrialist had moved to the powerful position of chairman of the board. But while Harriman's unrelenting, aggressive personality brought him success in the railroad world, his drive for power had proved physically exhausting. The family doctor, his close friend Lewis Rutherford Morris, ordered him to rest from business for several weeks on an extended vacation.

Harriman's vacation would not be an ordinary one. Accustomed to large-scale ventures and risks, he first decided on a big-game hunting trip akin to the Western adventures made famous by Theodore Roosevelt. Then his friend Daniel Elliot, the curator of Chicago's Field Columbian Museum, roused his interest in hunting Alaska's Kodiak bear with tales of the enormous size and strength of the creature. Having conquered the Union Pacific Railroad, Harriman set his sights on bagging a different sort of game. But it soon became apparent that his Alaska cruise would involve far more than a simple family outing. A hunting expedition into the Arctic wilderness also required luxurious accommodations for Harriman's family, and the railroad tycoon saw the opportunity for a still grander voyage. Capable of doing nothing in a small way, Harriman turned his hunting trip into a dazzling contribution to the world of science. In a single, sweeping gesture, a scientific expedition would build Harriman a public image as a philanthropist, perhaps even raise him to the stature of the sainted Carnegie.

The son of a clergyman, Harriman had ended his own education at fourteen when he insisted on entering the business world as a quotations boy on Wall Street. With an excellent memory and a "nose for money," he had risen rapidly in the financial district as a broker. Later, however, his lack of formal education made him an outsider in elite business circles. Harriman's obscure background created suspicion among the board members of the Union Pacific when he joined the company in 1897, and a contemporary of Harriman's wrote that "he was looked at askance, somewhat in the light of an intruder. His ways and manners jarred somewhat upon several of his new colleagues, and he was considered by some as not quite belonging in their class, from the point of view of position, financial standing, and achievement." Perhaps a desire to prove his status compelled him to pour his new money into the lavish expedition. For several weeks the railroad entrepreneur would have his own "floating university."

Alaska was an ideal choice for the expedition. Its coasts and waters abounded with uncatalogued species of plants and animals. While the Kodiak bear promised hunting trophies for Harriman, the rough wilderness virtually guaranteed new scientific discoveries. But the Arctic territory offered even more than scientific and hunting rewards. For the past decade, the nation's upper class had journeyed in ever-increasing numbers on fashionable sight-seeing tours to Alaska. An Alaskan cruise had become a badge of status, and the most up-to-date households proudly displayed souvenirs acquired on tours such as those sponsored by the Pacific Coast Steamship Company. Railroads thrived on such tourism, shuttling East Coast sightseers across the West's vast expanses to scenic parks and vacation spots. The Alaska expedition would promote interest in the tourist areas made accessible by the nation's railroads.

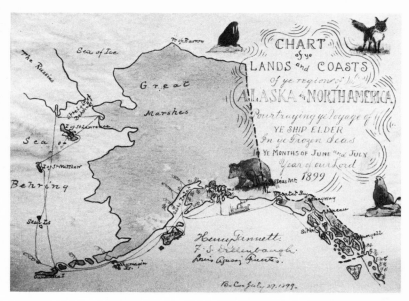

Chart of ye Lands and Coasts of ye regions of Alaska in North America, by Henry Gannett, F. S. Dellenbaugh, Louis Agassiz Fuertes.

Moreover Harriman, as president of the Union Pacific, was not unmindful of the potential inherent in any frontier. For two years the gold rush to the Klondike had competed on the pages of the nation's newspapers with the Spanish-American War. The financial pages told an even more enticing story. Alaska was a land of enormous economic potential—not only in minerals, but in timber, seal and walrus skins, ivory, and fish of all kinds. No one knew or cared to project the limitations of its economic potential, all of which could be exploited by cheap labor imported from the Orient as well as by "properly trained" native Americans. It was an uncharted province so vast that it competed favorably with that land once deeded by King Charles to Prince Rupert and now called Canada. To "seize the day," to be "in on the ground floor" of a bonanza far larger than mere gold hunting was a prospect to set before an energetic man of vision.

But the enterprising Harriman may have calculated still another purpose for his voyage. With a far loftier goal in mind than that of the typical wealthy tourist wanting to impress his friends with anecdotes of Alaskan travels, Harriman was beginning to visualize the most grandiose railroad scheme of all: a line that would circle the world. A railroad connecting Alaska to Siberia would be a first step in achieving his fantastic dream. A small rail line had recently been built in the Yukon goldrush country, with the incredible hope that it would eventually be extended beneath the Bering Sea to Siberia, and Harriman actually seems to have been eager to assess the potential for this daring operation to link the continents.

Only two months before he planned to leave for Alaska in May, Harriman began to accumulate the scientists he needed for his expedition. On a morning late in March of 1899, the railroad magnate paid an unannounced visit to Dr. C. Hart Merriam, the head of the Biological Survey of the U.S. Department of Agriculture. Harriman entered Merriam's office to solicit his help in organizing the expedition, but his wealth and power were so recently acquired that the scientist, far more familiar with the world of biological research than business, had never heard of his caller. The bewildered Merriam carefully humored Harriman, thinking him a harmless eccentric with an outrageous and amusing scheme. A quick check, however, confirmed Harriman's identity and that evening the two men met again to discuss plans for the expedition, which, as Harriman made clear, would be all at his expense. The distinguished scientist had no idea that he was about to embark on a project that would consume twelve years of his career.

The following day Merriam collected two of his colleagues, William Healey Dall, whose numerous trips to Alaska had made him the country's top expert on the region, and Grove Karl Gilbert, an eminent explorer-scientist of the American West, to meet with Harriman. Over the next few days ideas for the voyage began to take shape in several hurried meetings between the scientists and the railroad tycoon. On the last day of March a telegram from Harriman arrived at Merriam's office, inviting him and Dall to travel to New York in his private railroad car for yet another discussion. Over dinner at the elegant Metropolitan Club in downtown Manhattan, Harriman placed full responsibility for the expedition in the hands of the scientists, authorizing Merriam to invite the nation's most respected scholars. Impulsive, large schemes had catapulted Harriman to great success in the business world, but it was now up to the scientists' cautious, scholarly methods to complete the details that would make Harriman's vision a reality. Merriam suggested names of prominent scientists from New York to California, many of whom he knew personally from the elite Cosmos Club in Washington and from the Washington Academy of Sciences. Headed back to the capital on the midnight train, Merriam and Dall mulled over possible guest lists for the expedition.

After arriving in Washington early in the morning, Merriam turned immediately to the task at hand. The inner circles of Washington's Cosmos Club provided the nucleus of the expedition's party. Since its founding in 1878, the club had been a gathering place for the nation's top men of science, and Merriam, Dall, and Gilbert, the first scientists chosen for the voyage, often shared cigars and brandy in its plush rooms. Within its prestigious ranks an easy social ambience prevailed in which scientists traded stories of their adventures exploring the rough territories of the West. Merriam's first recruit for the expedition was Frederick V. Coville of the U.S. Department of Agriculture, a fellow club member who had earned the respect of other scientists with his many biological surveys in the West. In 1891 Coville and Merriam had weathered the hardships of an expedition to Death Valley together. Merriam had met another of his choices for the expedition, Henry Gannett, in 1871, when they both journeyed to Yellowstone with the F. V. Hayden expedition, which had introduced the photographer William H. Jackson and the painter of grandeur, Thomas Moran, to the wonders of the paint pots and spuming geysers of Yellowstone. That expedition had done more than anything else to persuade Congress to set aside Yellowstone as the world's first national park. So Gannett was well aware of the process whereby the value of unspoiled nature could be brought to the attention of the American public. He was one of the founders of the Cosmos Club, and as Chief Geographer of the U.S. Geological Survey, he had spent much of his life studying and living out in the rugged Rocky Mountain West. He readily accepted Merriam's invitation to explore Alaska. Also on the early Hayden expedition with Merriam and Gannett was a twenty-two-year-old ornithologist, Robert Ridgway, a gangling self-taught veteran of Clarence King's Fortieth-Parallel Survey across the West. Upon his return from that duty, Ridgway had developed a friendship with Merriam which had continued when they were both young assistant ornithologists at the Smithsonian Institution, living in the building's tower rooms. Now more settled, Ridgway confessed that he had "little love for the North," but agreed to join the Harriman expedition, taking leave from his position as Curator of Ornithology at the Smithsonian. Merriam's assistant at the U.S. Biological Survey, Albert K. Fisher, eagerly signed on next. Fisher had also been on the 1891 Death Valley expedition, and his growing reputation as an ornithologist would soon place him in the ranks of the nation's most accomplished scientists. Then William Trelease, the Director of the Missouri Botanical Garden, wired his acceptance from St. Louis. Drawing luminaries from each scientific discipline, Merriam carefully hand-picked his crew. Only the best would do.

However, Merriam stepped outside the world of organized science when he contacted the naturalist John Muir in California. Muir's long treks through the Sierras and the Arctic country had instilled in him a great respect for the wilderness, and his writings staunchly advocated preservation. Thus Merriam's offer of a free trip to Alaska met with suspicion from the West Coast naturalist. The name Edward H. Harriman was unfamiliar to Muir, and, reluctant to become indebted in any way to a tycoon, he demanded the full details about the nature of the expedition. Only after Merriam explained that the voyage would explore areas of the Alaskan coast that even

the adventurous, well-traveled Muir had not seen did the seasoned explorer agree to sign on. Muir was not the only guest
whose scientific credentials did not conform to the academic
mold. When the pastoral author John Burroughs agreed to
serve as the expedition's historian, he lent a sentimental appreciation of nature to the fact-finding mission of the scientists.
Burroughs' many nature books had struck a receptive chord in
the American public, and in 1899 he was a popular national
figure. With a flowing white beard and poetic sensitivities akin
to those of his friend Walt Whitman, Burroughs had endeared
himself to his national audience, and even the rugged Theodore Roosevelt listed Burroughs' gentle works as among his
favorites.

After securing Burroughs' acceptance, Merriam returned to
the academic world of science for his recruits. At Yale, Professor William H. Brewer read Merriam's letter of invitation
with excitement and surprise. At seventy-one, the gray-
bearded, bespectacled botanist was an old hand at Western
exploration as a veteran of the California Geological Surveys
of the 1860s, when he had pioneered forays into the remote
Sierran wilderness. Brewer also knew well the risks involved
in an Arctic voyage such as Harriman planned to undertake.
Several years earlier he had journeyed to Greenland on a
steamship that struck an iceberg as it approached the Arctic
Circle. Perhaps the old explorer saw the Alaska expedition as
his last chance at adventure, and he wired his acceptance back
to Washington.

For the expedition's expert in geology, Merriam turned to
Professor Benjamin K. Emerson at Amherst College in Massachusetts. Emerson's positive reply to the proposal was quick,
and by the fourth of April he had signed on as one of Harriman's guests. After warning Emerson that Harriman wanted
the expedition plans kept confidential until close to the departure date, Merriam enthusiastically wrote the professor that
"your college authorities must not look on this expedition as a
'junketing trip' or vacation. On the contrary, it is a phenomenal opportunity to obtain material and illustrations for future

work and lectures." In turn, Emerson invited his young protégé, Charles Palache, a mineralogist at Harvard who, at thirty,
had already pioneered in the field of seismology, discovering
the first evidence of geological fault lines when he mapped the
San Francisco Bay Area a few years earlier. Though thrilled
at the prospect of exploring Alaska, Palache hesitated. His
wedding date had been set for June 21, when the expedition
would be steaming among remote Arctic glaciers. A talk with
his young fiancée, however, convinced him that the wedding
could be postponed, and Palache's name was added to the
growing roster of expedition members.

George Bird Grinnell, with whom Merriam had hiked the
rocky slopes of Mount Rainier in 1897, had recently published
two volumes of Indian lore, and Merriam felt that his expertise on the Alaskan natives would be a valuable contribution
to the expedition. And together, the two scientists recalled the
name of Edward S. Curtis in Seattle when Harriman requested an official photographer for the voyage.

Photography would not be the only means of "capturing"
the Arctic scenes for posterity. Harriman invited three artists
along on the trip. Perhaps stirred by Theodore Roosevelt's
grand tales of the West, Harriman sought out one of the illus-
trators of Roosevelt's early hunting books, Robert Swain Gifford. By 1899, Gifford's landscapes enjoyed an admiring
audience in New York, and he was a highly respected art
teacher at The Cooper Union. Nearly sixty years of age, Gifford
collected his old adventurous spirits when Harriman wired him
an invitation to participate in the expedition. Gifford's colleague on the voyage would be Frederick S. Dellenbaugh, who
had been an artist on the Powell expedition down the Colorado River in 1871–1872. As a teenager on the Powell expedition, Dellenbaugh had combined his topographical skills with
his artistic ability, and, like Gifford, he had not lost, in the
intervening years, his urge to explore new territory. On the
Powell expedition, Dellenbaugh had helped to make the first
maps of the Grand Canyon area, and he had journeyed across
the rugged Southwest on horseback to deliver the maps to Salt

Lake City. The two artists excitedly traded letters about the upcoming Alaska expedition, and Gifford wrote that "we can do a great deal together when the other specialists are entirely absorbed in their respective pursuits. On such a trip there should be, if possible, 'two of a kind.'" Indeed, Dellenbaugh and Gifford, both old veterans of western travel in their searches for impressive artistic subjects, seemed "two of a kind."

For the expedition's third artist, twenty-five-year-old Louis Agassiz Fuertes, the voyage to Alaska would be a first brush with the world of adventure and exploration. The young artist, whose bird paintings had already met with critical acclaim, signed on the expedition as the ornithologist Albert Fisher's assistant. Fuertes' fatherly mentor, the famous ornithologist Elliott Coues, fretted about his protégé's upcoming trip, writing him, "I want to impress upon you the necessity of taking the best possible care of your health. You will have plenty of beastly weather—and often be much exposed to the wet. Look out for that—you cannot get too well provided with clothing—both underclothes for a dry change, and waterproof outside. Also look out for *accidents*." Coues, who died a few months after the expedition's return, urged Fuertes to come back to the East Coast "with very full portfolios of all sorts of interesting things, as well as in perfect health, ready to resume and push to completion the important business you have on hand with my publisher and myself." Coues had offered his young student the opportunity to illustrate the new *Key to North American Birds*, and he worried that the Alaska expedition would delay its publication.

In Merriam's Washington offices acceptances came from scientists across the country, and the list of guests continued to grow. The poet Charles Keeler, an outspoken Director of the Museum of the California Academy of Sciences, wired his acceptance from San Francisco, and his colleague William E. Ritter, the President of the California Academy of Sciences and a zoology professor at the University of California at Berkeley, signed on in late April. A few days later, Cornell

University's Dean of the Forestry School, Bernhard E. Fernow, wired Merriam a positive reply. Only a month remained before the expedition would be launched, and Merriam enthusiastically noted in his journal that the scientists were eager to leave.

Excitement over the trip ran high, and it was inevitable that, despite Harriman's orders to the contrary, word of the expedition leaked out to the press. The *New York Herald* scooped the story in late April, splashing a headline at the top of a page that hailed Harriman's distinguished scientific party and their "invasion of Alaska." Merriam breathed a sigh of relief that the announcement had come from New York and not Washington. The ensuing publicity frustrated Harriman's penchant for privacy, and people eager to participate in the expedition deluged his offices with mail. A tighter rein on information was kept in the following weeks, and Harriman gave even the expedition members only the most essential facts regarding the date of departure and the schedule. An air of secrecy surrounded the letters and telegrams that circulated among the elite network of expedition guests, and when the date of departure was set from New York for May 23, Gifford warned his fellow artist Dellenbaugh against mentioning word of the date to the press.

When the final list of expedition guests had been drawn up, Harriman found that it included twenty-three of the country's top scientists representing twelve fields, three artists, two photographers, two physicians, two taxidermists, and one chaplain. In the span of a single month Merriam had pulled together a group of eminent professionals and had secured the Washington Academy of Sciences' formal cooperation with the expedition. Only the details of transportation and equipment remained to be settled. On the West Coast, Harriman ordered the Oregon Railroad and Navigation Company to refurbish the *George W. Elder*, an old iron steamship, into a scientific luxury liner. The East Coast expedition members would travel on Harriman's Pullman parlor cars on a sight-seeing trip across the country before they boarded the ship in Seattle.

In part due to the secrecy he maintained, and in part due to his own fast-rising reputation as an empire builder, Harriman's spectacular scientific "vacation" portended something more than a holiday to the celebrity-watchers and financiers of the late Gilded Age. No philanthropist, or government bureau for that matter, up to this time had ever assembled such a group of scientific experts and set them on such a mission. Surely pure knowledge was not the objective. Rather, it must be something of a practical nature, something large, grand in scope or vision, as befitted the age that had produced the Suez Canal, the Eiffel Tower, the transcontinental railroad, towering skyscrapers served by Otis's elevators, the telephone-talking-wire, Brooklyn's stupendous swinging bridge, Panama's Canal—then in progress—and the dynamo-powered World's Columbian Exposition in Chicago, which only fifty years before had been a rude frontier town like Juneau or Skagway. Perhaps a merger of all western railroads was in the works; perhaps a whole new route to Alaska and the Arctic frontier. Maybe Harriman intended to purchase Alaska itself and then build his railroad around the world. The idea seemed preposterous, yet nothing was impossible in the era of sophisticated finance and high technology, both of which E. H. Harriman epitomized.

CHAPTER TWO

ACROSS THE CONTINENT
IN PULLMAN'S PALACE CARS

Beneath gathering rain clouds on the morning of May 23, the East Coast expedition members hurried about, making last-minute preparations for the train departure from New York at two o'clock that afternoon. In New Haven William Brewer rose at five-thirty, said good-bye to his family, and headed to the train station to make the Harriman connection on time. In Manhattan Frederick Dellenbaugh busily ran errands that had piled up until the day of departure. The artist had a sketching umbrella repaired and checked his mail at the Century Club, where a friend pressed upon him a bottle of coca leaf extract, with the assurance that it would guard against fatigue and seasickness. Dellenbaugh may have been apprehensive about the trip: he had signed his will three days earlier. At the Century Club Dellenbaugh met his fellow artist Swain Gifford, and together they proceeded to Grand Central Station to meet the rest of the party, assembled to board the train that would carry them to the West Coast in high style. The two artists found a quiet, sedate group of passengers nonchalantly awaiting Harriman's private train. When Harriman strolled toward his guests, Dellenbaugh was struck by the feeling that such a grand tour was a mundane matter for the railroad king. "Mr. Harriman soon walked in as unconcernedly as if this sort of princely tour were an everyday matter with him," he wrote in his diary.

A luxurious train of "palace cars" rolled to a halt on the tracks. Since no passes had been issued the elite group, the guests had only to say to the gateman "special train," as if the words were somehow passports into the powerful world belonging to Harriman. Once inside, the expedition members concealed their awe at the plush surroundings, as they surveyed the five cars Harriman had made available for the week-long trip. In the smoking car, which Harriman had appropriately named the "Utopia," the scientists found wicker settees and armchairs, an entire library of Alaska books, and several brands of the finest cigars. From the wide, sparkling windows of the "Utopia," Harriman's guests could watch the passing spring landscape while they engaged in small talk.

Once his guests had settled themselves, Harriman made his way through the train to meet the passengers. One by one, the men rose to shake the railroad baron's hand. But Harri-

Traveling in "Utopia": The Harriman Train.

man's calm, collected presence and his clear, penetrating eyes unnerved several of the scientists as they anxiously greeted their host. Eager to make a good impression, some of the men jumped up so quickly from their seats to meet Harriman that they knocked their heads against the sleeping berths above. After several such embarrassing accidents Harriman laughingly reassured a flustered Thomas Kearney, a Department of Agriculture botanist, that he was the fifth to bump his head upon meeting him.

Sitting alone in a window seat, John Burroughs already felt homesick as the train rolled past his peaceful farm and vineyards on the Hudson River in the late afternoon, and he thought he glimpsed his wife waving her white apron. Burroughs sadly pulled out his journal and wrote, "Have I made a mistake in joining this crowd for so long a trip? Can I see nature under such conditions?" Never a great socializer, the naturalist found his scientific company full of strange jargon, and he kept quietly to himself, brooding over his decision to spend two months away from his treasured vineyards. "But I am in for it," he concluded, and with that worried entry, he tucked away his journal for the day and watched the passing scenery.

Meanwhile, Frederick Dellenbaugh was having the time of his life exploring the fancy gadgetry and plumbing in the sleeping car he was to share with Louis Fuertes. The washstand, he found, was full of surprises. When you pulled a handle the water basin appeared, and pulling yet another handle produced a jet of water, either hot or cold. Pushing a button yielded cool drinking water. Dellenbaugh decided that, altogether, it was a quite satisfying arrangement.

Harriman, always thinking of efficiency and organization, mulled over the various committees that he would form within the expedition party. By the time the train had reached Chicago the following day, he had carefully assigned his guests to serve on groups ranging from an Executive Committee, which he himself headed, to a nine member committee on literature and art. Though Harriman repeatedly stressed that he would not interfere with the scientists' independent work, he set about applying a corporate model of organization to the group, stepping easily into his usual authoritative role. C. Hart Merriam later wrote that Harriman's method was one that "perfected its plans in advance and took advantage of every opportunity for work." Typically, the railroad magnate had not wasted a day before he put his efficiency plans into effect.

Some of the younger members of the expedition, however, created a lighter mood in the rail cars. Piecing together words from their old college songs, Fuertes, Palache, Coville, Fernow, and the physician Edward Trudeau passed the time by serenading the teenaged Harriman daughters, Mary and Cornelia, along with their cousin Elizabeth Averell and their young friend Dorothea Draper. As the train chugged through the Rocky Mountains, joviality and excitement filled its passengers. Even John Burroughs emerged from his ponderings long enough to take a joyride on the platform jutting out before the train engine, called the "cow-catcher." Wind-blown but exhilarated, the old naturalist heartily recommended the ride to his colleagues, and soon scientists filled the cow-catcher, straining against the fierce wind. From the vantage point offered at the engine, Burroughs had been struck by the erosion of western land by the railroads and he shuddered at the destruction of nature. Later, he noted in the expedition's official history, that "in places the country looks as if all the railroad forces of the world might have been turned loose to delve and rend and pile in some mad, insane folly and debauch." It was, of course, the same force that had brought Harriman the wealth to sponsor the expedition. Oddly enough, Burroughs did not seem to imply irony in his observation.

Coal mines had also stripped the western scenery of its natural beauty. After passing Cheyenne, which Dellenbaugh noted was no longer the crude frontier town with one wooden hotel and station that he had visited nearly thirty years ago, the train creaked to a halt. Here at Bitter Creek a desolate little community had risen at a large coal mine. Walter Devereux, the expedition's mining engineer, shared his expertise with the rest of the party. His descriptions of the smelting operations that he had recently set up in England left some members of his audience amazed that Americans were now teaching the British about such technological matters. As the train pulled away from the coal mines toward the Wind River Mountains, Dellenbaugh eagerly craned his neck for a glimpse of the small cove on the Green River where twenty-eight years ago he had embarked on John Wesley Powell's famous second expedition to explore the Colorado River area. The train sped past, but the artist spotted the old launching site clearly enough to note the little shanties that had sprung up in the Green River community. The Old West, he noticed, had changed drastically in the intervening years, and frontier towns that once had boasted a single saloon now raised numerous buildings against the western horizon. Railroads linked the far-flung communities to the rest of the nation, and their coal now supplied energy to the urban regions along the East Coast.

Once the domain of the lonely cowboy and the isolated cattle baron, the West now belonged to the railroad industrialists, who raced across its rugged terrain in their private palace cars, continually surveying the endless possibilities for development. When the expedition had passed Omaha, the Union Pacific's president, Horace G. Burt, joined his private car to Harriman's rail caravan for the ride into Wyoming, and the two Union Pacific rulers conferred about upcoming business deals in the privacy of their own plush office on rails.

The world of Harriman and Burt was a remote one to most of the scientists, who spent the week-long trip absorbed in their own observations of the West. Fernow and Coville, both botanists, argued heatedly about whether Fernow had actually spotted a'lodge pole pine, while C. Hart Merriam, well known for his salty stories and wry sense of humor, entertained his companions with tales of his explorations in the Rockies. The train passed mountain ranges which Merriam had personally named in his government surveys years before, and the region still held its old fascination for the biologist. Louis Fuertes was amazed at the rich experience and travel claimed by his prestigious older companions. For both the young bird artist and his new friend John Burroughs, the trip through the West was a first-time journey into strange country, and the two often sat together on the train, marveling at the vast stretches of wilderness and pointing out birds to each other. Fuertes wrote his mother that he and Burroughs were like two young children in their delight in the countryside. By the end of the expedition, Fuertes was affectionately referring to the old naturalist as "Uncle John."

"Uncle John" Burroughs, however, was busy making continental comparisons in his mind and in his notebook, contrasting the vast giant-step landscape of the West with cosy, comfortable New England. The Great Plains affected him "like a nightmare." Gone was "the sheltering arm of the near horizon about us." Instead, "A night's run west of Omaha a change comes over the spirit of nature's dream." The absence of trees seemed to make the landscape "youthful," or as he put it in one of his many Whitmanesque similes, "like the face of a beardless boy." Sometimes in true Whitman fashion he could not make up his mind whether the West represented youthful masculinity or whether it was really "feminine." Throughout the journey west "Uncle John" seemed constantly bound up in sexual reveries that expressed themselves in metaphorical characterizations of the landscape. Across the Atlantic, Sigmund Freud had begun to think of this as Victorian "sublimation."

Still, his observations were striking. "Before we get out of Wyoming," he wrote, "this youthfulness of nature gives place to more newness—raw, turbulent, forbidding, almost chaotic. The landscape suggests the dumping ground of creation." Scale also stunned him: "What one sees at home in a clay bank by the roadside on a scale of a few feet, he sees here on a scale of hundreds of thousands of feet—the erosions and sculpturing of a continent, vast, titanic mountain ranges like some newly piled earth from some globe-piercing mine shaft, all furrowed and carved by the elements, as if in yesterday's rainfall."

When they traversed the Badlands country he fantasized,

"The earth seems to have been flayed alive . . . no skin or turf of verdure or vegetable mould anywhere, all raw and quivering." The dramatically eroded hills looked to him "as new and red as butcher's meat, the strata almost bleeding." Had he been better informed, he would have been astounded at the fact that these Badlands had indeed been the site of bleeding—millions of years ago when they became a vast dinosaur graveyard. In this sense the West was not "new" but very "old." For the paleontologist it had become a window into the earth's antediluvian past, which had captured the imagination of the world as Professors O. C. Marsh and E. D. Cope hauled trainloads of dinosaur bones back to their "cosy" eastern laboratories. But Burroughs, like most Americans influenced by the antiquity of Europe and the ancient East, tended to read the history of America from east to west. He focused on the new settlers in raw unfinished towns set down as if by accident on a landscape not yet completed. He was not unique in this attitude. For much of the time before the Civil War, the Great Plains and the Badlands at the foot of the towering, unreal Rockies were in the public mind and even the scientific mind the "Great American Desert," or, as that arch-romantic John C. Fremont put it, "the Great Zahara of North America."

But Mr. Harriman's mechanical caravan rolled on, and once in the Rockies, the picturesque and even the sublime began to reassert itself. On the journey's fourth day the train drew close to Shoshone Falls in Idaho, and Merriam's eyes began to shine with a new idea. He remarked to Harriman that the twenty-five-mile trip to the falls would provide a pleasant break from the rail cars, if only transportation to the river could be arranged. Hearing a challenge, Harriman immediately wired to Utah, ordering horses, a stagecoach, and two buggies to be brought up by rail. Nothing seemed impossible in the world of the railroad baron, and the scientists marveled at the ease with which Harriman exercised his power. The party set off through the Idaho countryside early in the morning, most galloping along on horseback while the wagons lumbered across the rolling plains. Six horses pulled the bright red stagecoach, whose weathered old driver had headed coaches throughout the West for forty years; the trip to Shoshone Falls was one of the least dangerous in his long experience. To the north, the snow-covered mountain ranges loomed ahead of the Harriman caravan as the expedition moved through fields covered with the wildflowers of late spring. Sitting up on the stagecoach, Albert Fisher, Merriam, Gilbert, Fernow, Ridgway, and Dellenbaugh exchanged stories of their past adventures and chatted in scientific language about the land around them. An ecstatic Robert Ridgway spotted species of birds that he had not seen in thirty years.

At the Snake River Canyon the party drew to a halt to survey the first glimpse of Shoshone Falls before they made their way down the path to the foot of the waterfall. A boat for tourists ferried them across to the opposite side of the Snake River for the best view of the falls. There a small wooden hotel stood, its eaves dripping with thick mist from the rushing falls nearby. A wizened old guide led the party farther down into the canyon for the view from the bottom. They clambered down wooden ladders placed for tourists against the side of the canyon. At Shoshone Falls, Burroughs felt more in his Ruskinian element. He found the falls "a sudden vision of elemental grandeur and power opening at our feet. The grand, the terrible, the sublime are sprung upon us in a twinkling." With the sublime and the picturesque all about them, it was only appropriate that the expeditioners hold that pastoral rite— a picnic. In ecstasy Burroughs declared, "Baptize the savage sagebrush plain with water and it becomes a Christian orchard. . . ."

After the dreamlike picnic, the group mounted the horses and wagons for the return trip to reality, while Harriman and Merriam traveled on to find the Blue Lakes, below the falls. Tired and sore from their horseback or wagon rides, the scientists arrived back at the station eager for a mug of beer and a hearty dinner. Dellenbaugh claimed the beer was the best he'd ever tasted, and he carefully noted the brand, Pabst Blue

Ribbon, in his journal before he turned in for the night.

The train arrived in Boise just as the sun rose, and Harriman's party found the town's chamber of commerce officially welcoming them at that early hour. The local newspaper hailed Boise's distinguished visitors and announced to its readers that Harriman was "the man of the hour in railroad circles." His arrival in town became a grand cause for celebration. Town leaders escorted the entire party onto trolley cars covered with American flags for the occasion and paraded the group down to the Boise Natatorium, where they bathed in warm spring waters. Later, the visitors talked enthusiastically with Boise's businessmen of the opportunities for the town's growth. Ores from the surrounding mountains and irrigation techniques promised prosperity for the community. But Harriman's schedule prevented a long conference, and before noon the train was already puffing its way through the Blue Mountains— near the old Oregon Trail—leaving behind an awe-struck group of Boise citizens.

Before leaving Idaho, Harriman planned yet another excursion for his guests. This time they would journey down the Snake River for several miles on a stern-wheel steamer. A. L. Mohler, the president of the Oregon Railroad and Navigation Company, which Harriman had recently acquired through the Union Pacific, had invited the party to cruise briefly down the river on his boat. A special train from J. P. Morgan's Northern Pacific line (which, two years hence, Harriman would make a daring move to control, causing the Northern Pacific panic of 1901), carried the party through the Idaho hills to Lewiston, where they boarded Mohler's steamer. A gentlemanly sense of hospitality prevailed between the opposing rail barons, as the president of the Northern Pacific escorted his rival's party in their brief journey along his line. Temporarily at least, the ruthless scheming that took place behind the closed doors of corporate boardrooms was replaced by the sense of aristocratic camaraderie that bound the powerful competitors in the same elite world.

The Northern Pacific train chugged through the Nez Percé

Indian Reservation. Coville and Fernow, completely unaware of such things as railroad rivalries, puzzled over why the pine trees were seeding so rapidly in the hills. The two botanists finally concluded that the prairie fires that once destroyed the young trees along the hills must have been brought under control, allowing the saplings to cover the countryside. They were deep in serious theorizing when a sudden jolt brought their conversation to an abrupt halt. The train's wheels screeched to a stop as curious scientists scrambled to the windows. A derailed car blocked the line and workers were madly trying to hoist it back onto the tracks. Embarrassed by the delay along his lines, the Northern Pacific executive angrily ordered that the workers simply drop the stray car into a ravine below the tracks rather than keep the important Harriman waiting. The workers, however, quickly got the car back in its proper place, preventing needless wreckage. While the train waited, Dellenbaugh took advantage of the delay. He pulled out his sketching pad and quickly drew the two squaws on horseback who had ridden up to investigate the accident.

At Lewiston, the expeditioners climbed aboard a large stern-wheeler and swung away down the Snake River's rapid current. To get a better view of the countryside, several men climbed up onto the hurricane deck, ignoring a large sign that warned passengers not to trespass on deck. A certain headiness pervaded the party and Dellenbaugh noted in his journal that "signs mean nothing to our party. We own everything!" In Harriman's world, one made one's own rules.

In fact, the trip down the Snake was more than a mere pleasure jaunt for the tycoon. One of his rival rail lines was building a new railroad along the river and, together with his colleague Mohler, Harriman wanted to assess the grading along the riverbank. Railroads had already replaced the steam transportation lines down that section of the river, and the steamer's excursion past its usual stop at the railroad was a rare one—so rare, in fact, that the boat's captain was completely unfamiliar with the river. In the darkness, he nearly smashed the steamer into the buttresses of a bridge. But after careful manuevering, the boat finally bumped the shoreline, and its relieved passengers scrambled off. Harriman's special train waited for them at the site, and they were soon sipping whiskey in the comfort of "Utopia."

A brief stop to admire the Snake River's Multnomah Falls found John Burroughs once again entranced with the sensuality of nature. The falls, he fantasized, were like a beautiful but coy woman, a "nymph" who had "withdrawn into her bower, but had left the door open." Ever the voyeur, he found "this element of mystery and shyness well nigh irresistible." He lusted in the pages of his journal: "How the siren mocked us, and made the few minutes in which we were allowed to view her so tantalizingly brief." But the train was bound for Portland, and its schedule prevented the dreamy Burroughs from resting at the foot of the falls as long as he might have liked. Even after viewing the spectacular scenery of Alaska, Burroughs wrote that nothing diminished his memory of those "goddess-like" falls.

While Harriman's train roared toward the West Coast, another train hurried its passengers along the Pacific Coast toward Portland from San Francisco. On it, John Muir and Charles Keeler shared a sleeping car on the three-day journey to meet the Harriman party in Portland. Muir entertained his new friend, whom he found a "charming companion," with tales of the Alaskan wilderness he had so often explored. But Keeler's mind was already on the family he was leaving farther and farther behind. He had joined the expedition in hopes of making some extra money from the articles he could write about the journey, but he did not look forward to the two months away from his wife and young daughter. In fact, Muir also had not been particularly anxious to leave. The day before his departure from San Francisco, the explorer had written his editor, Walter Hines Page: "I start tomorrow on a two months' trip with Harriman's Alaska expedition. John Burroughs and Professor Brewer and a whole lot of good naturalists are going. But I would not have gone, however tempting, were it not to visit the only part of the coast I have not seen and one of the scenes that I would have to visit sometime anyhow." Muir's year had been spent battling powerful ranching interests to preserve the forests from the grazing of sheep and cattle, and he was weary from his tedious efforts to win politicians to his cause. The garrulous Muir was full of talk as the train made its way along the coast, and Keeler heard stories of Muir's recent political fights as well as details of the naturalist's background. Muir gave his cabinmate little time for homesickness, chattering about his boyhood in Scotland and his tyrannical father. Perhaps sensing Keeler's reluctance to travel too far from his family, Muir pointed out the various species of trees the train passed, and both men sadly observed the destruction of the forests of northern California and Oregon by the lumber industry. The government, Muir argued, must take immediate steps to protect the nation's forests from the insatiable demands of industry. He was almost alone, however, in this "radical" point of view.

Muir disdainfully eyed the tourists on board the train, as they clustered in passive groups to look at the sights their guides pointed out. He sneered that "people look at what they are told to look at or at what has been named. Nameless things, however fine, go unnoticed." But Muir took it upon himself to ensure that Keeler missed nothing of the surrounding countryside. Keeler wrote his wife that Muir "is almost as great a talker as Aunt Sophie so you can imagine I don't have much time to myself." Nevertheless, the young San Francisco poet found time to chat with a fellow passenger who was headed to Portland to meet with Harriman on business matters. From the businessman, Keeler discovered for the first time the magnitude of Harriman's power in the railroad world. His new acquaintance, he found, was also a friend of the publishing tycoon S. S. McClure, the owner of the popular *McClure's Magazine*. Keeler learned that the powerful McClure had secretly bought out Harper's Publishing Company. Being privy to such inside information in the national business network was new to the naïve Keeler, and he was beginning to understand the elite circles he was soon to join.

In Portland Muir and Keeler checked into the lavish Portland Hotel to wait for the Harriman party, due in town the following day. The outspoken Muir cornered several of his influential friends in the lobby of the hotel and continued his tireless, unending efforts to garner support for forest protection. In no uncertain terms, Muir informed the president of the Mazamas, a local mountain-climbing club, that time was running short for the preservation of wilderness. President Steel of the Mazamas promised to oppose the pasturing of sheep in Rainier Park and in the Cascade Reservation, but two other town leaders, Judge George and a man named Hawkins, whom Muir described as "fat and easy," were noncommital on Muir's favorite subject. While Muir spent his time arguing for his lifetime cause, Keeler wistfully thought of his wife. The extra day in Portland was wasted time, he thought, and he could have spent it in San Francisco with his family. After eating an elaborate dinner, accompanied by a large orchestra, in the dining hall of the hotel, Keeler wrote home of the unaccustomed luxury around him, and enclosed a newspaper clipping that reassured his wife of the lavish provisions for the expedition.

On May 30, amid great secrecy to avoid the press, the Harriman train rolled into Portland. Harriman had arranged for a trolley car to take his guests to the top of Portland Heights, a hill from which the local countryside could be seen. But it was Decoration Day and the town's populace had come out to celebrate in great numbers, streaming onto Harriman's trolley car, unaware that it was a private car. After guards escorted the uninvited passengers off the trolley, the car climbed the hillside, only to find that rain and fog obscured the view. Therefore, it was not long before the trolley returned and deposited the Harriman party at the Portland Hotel, where they met Muir and Keeler. To Muir's dismay, his old acquaintance John Burroughs grabbed him by the hand and immediately began introducing him to the scientists. Muir carefully disengaged himself from his exuberant friend.

Despite Harriman's efforts, stories of the expedition filled the Portland newspapers, hailing Harriman's voyage as "a striking illustration of the beneficent possibilities of great wealth. . . . Mr. Harriman," wrote the *Oregonian*, "has done his country and the cause of human learning a signal service." With a glowing fanfare the Mazama "mountaineers" escorted the Harriman group to a fancy new steamer recently acquired by A. L. Mohler. Said to be the fastest stern-wheel steamer in the world, the boat was to take the party down the Willamette River to Kalama, where Harriman's private train again waited to carry them to Seattle. There, the group would finally board the *George W. Elder* to begin their Alaskan adventures.

Along the route down the Willamette, other boats whistled their respectful salutes as Mohler's sleek steamer sped by. Merriam recorded the speed of the steamer at a breathtaking twenty-seven miles per hour. The brisk pace carried them swiftly to Kalama, where the familiar special train waited on the Northern Pacific tracks, which J. P. Morgan had courteously cleared of all regular cars: his rival Harriman would have straight, uninterrupted passage to Seattle. On board the train for the first time, Muir and Keeler met their host and both men found him surprisingly simple and unassuming.

Meanwhile, Edward Curtis waited anxiously in Seattle for the party to arrive. When the train pulled into its final destination, Merriam immediately sought out the photographer. Together the two men joined the rest of the party at the prestigious Rainier Club for an elaborate lunch. It must have been an impressive affair for the young Curtis, whose brush with "society" groups was normally limited to the wealthy Seattle citizens whose weddings and balls he photographed. Little did he suspect, as he feasted at the Rainier Club that day, that his entry into Harriman's elite world would mark a turning point in his career.

After lunch, Muir found himself waylaid by a reporter and grudgingly granted an interview to the journalist, who wrote an article on Burroughs and Muir entitled "Two Famous Men," a description Muir modestly disclaimed. While the naturalists talked with the newspaperman, their colleagues hurried about, transferring their baggage and equipment to the *Elder* and checking for last-minute mail from home at the Seattle post office. The last errands completed, the party assembled eagerly on board the ship, and crowds of well-wishers on the wharf shielded themselves against a drizzling rain to scrutinize the important passengers. But preparations took longer than expected, and after a four-hour delay the crowds had substantially thinned. Muir and Keeler took the opportunity to write hurried letters to their wives, convinced that this could be the last chance to send mail for several days. At six o'clock in the evening the *Elder* finally pulled away from the dock. Strains of music from the ship's graphophone drifted across the water to the last remaining bystanders on the rain-swept wharf.

CHAPTER THREE

"THE BEST OBJECT LESSONS
TO BE FOUND
ON THE COAST"

Unpacking in their staterooms, the expedition guests made themselves comfortable in the small quarters that would be their home for two months. Charles Keeler was writing his wife the third letter of the day when John Muir came tapping at his door. There had, it seemed, been an unforeseen shortage of rooms and the ship's pilot, Mr. Jordan, had appropriated Muir's own stateroom. Bags in hand, the seasoned old naturalist had come to share his new friend's room and together they carefully rearranged the baggage to accommodate two men in the cramped quarters. Keeler optimistically continued his letter home after the interruption: "Of course we are a trifle crowded but how great a privilege to be cooped up for two months in a little room with John Muir!" It would not be long, however, before the young poet would yearn for the privacy of his own home in San Francisco. But Muir was delighted with his new room. His old friend William Brewer was just next door, and close by was the cabin of John Burroughs, who, in the weeks to come, would become the favorite target of Muir's witty barbs.

The Good Ship *George W. Elder*, by E. H. Harriman.

The expedition members, though tired after a long day awaiting the ship's departure, were in high spirits over the upcoming cruise in Alaskan waters. They circulated freely over the spacious decks of the *George W. Elder* as the ship made its way toward Vancouver Island, where, on the first of June, it would make its first scheduled stop. That evening aboard ship, Muir discovered he held a position on Harriman's Executive Committee, an appointment which he assumed would give him a voice in the expedition's schedule and routes.

When the scientists woke the next morning, most of them saw only a wharf from their cabin windows—during the night, the ship had anchored at Victoria, British Columbia. Harriman had arranged for a special trolley to transport his guests to the Victoria Museum. There they browsed through Indian artifacts and stuffed animals from the Arctic region, seeing the creatures they were soon to view firsthand. Time allowed only a short visit, and soon the expedition headed northward again, through the channels between the region's islands and the mainland. Dall felt especially happy as the ship pulled away from Vancouver Island: in the ice at the landing site the Alaskan expert had discovered a mass of what he and George Bird Grinnell guessed was mammoth dung. Together they had loaded the precious prehistoric treasure on board the ship. Already, the expedition had gathered its first "specimen" for later study in Washington.

The *Elder* cruised slowly through idyllic waters in the Gulf of Georgia, where an extraordinary sunset drew the passengers to the deck en masse. An orange sun sank beneath the brilliant horizon, and the awestricken scientists clustered at the rail, hushed at the sight. But William Dall was not so easily impressed. The best scenery, he knew, lay still farther to the north, and he interrupted his companions' rapt silence to summon them to the lecture hall for a talk on his numerous experiences in Alaska. Most of the audience retired soon after Dall's scholarly lecture, amid rumors that the ship would traverse the often treacherous Seymour's Narrows early in the morning. Not wanting to miss this possibly dangerous section of the journey, most of the scientists left word that they were to be awakened at four, when the *Elder* would approach the narrows. Their hopes for adventure, however, were quashed; William Brewer, with his usual scientific precision, noted at five a.m. that a high tide had completely submerged the sharp rocks. The *Elder* plunged safely on into the deep Pacific.

All day the ship rocked in the swells of the open ocean, forcing several expedition members to leave their dinners untouched. One by one, they retired carefully to their rooms in an attempt to maintain their dignity. As the vessel bounced through the high waves, Dellenbaugh discovered that the ship had once enjoyed the dubious title of the "George W. Roller," a reference to the ship's tendency to roll and pitch. Though bilge keels had been added to the vessel when it was overhauled for Harriman, it still rolled badly with the ocean's whims. Indeed, in the weeks ahead, John Burroughs would spend much of his time suffering the effects of seasickness, and the sensitive naturalist was one of the first to seek his room that day. Before long, however, the ship left the rough

"The Admiral," Roland Harriman.

Lifeboat Drill. E. Harriman holding Roland Harriman (center); Averell Harriman (in large hat) to Roland's right.

Lifeboat Drill.

Captain Doran.

ocean for the more tranquil waters of the northwest channels, bound for Princess Royal Island.

Along the way, Charles Keeler pulled out some drawings by his wife, Louise. He had carefully packed them in his bags, and now he proudly showed the sketches to Swain Gifford for a professional artist's opinion. Gifford's polite enthusiasm delighted Keeler and he promptly shared the encouragement in a letter to his wife. Months later, Louise Keeler's sketches, drawn from photographs of the trip, impressed Merriam so much that he used her drawings to illustrate the volumes on the expedition that Harriman published. Such exchanges filled the early days of the luxurious voyage, and the scientists befriended each other with idle chatter and casual story-telling. William Brewer and John Muir competed with each other to see who could tell the most entertaining stories. Perhaps a bit uncomfortable on an expedition that could afford the amenities of the *Elder*, the scientists exchanged banter to alleviate their uneasiness. The original inhabitants of the region were a popular topic of conversation and a source of great curiosity to the entire party. At Lowe Inlet on Princess Royal Island, they had a chance to see the natives firsthand when the ship stopped to examine the small community of Indians and Chinese who worked in the town's salmon cannery.

Gilbert, more eager to make contact with raw nature than raw fish on the cannery assembly line, suggested that the scientists hike through the woods to investigate some glacial basins he had spotted from the ship's deck. Anxious to see if all the basins held the beautiful clear lakes that he had observed on the west side of the island, the tall, gaunt Gilbert led the way through the dense undergrowth. Even for the veteran hiker Muir, the thick forests of hemlock and cedar made movement slow and tedious, and the hike proved to be not worth their efforts. When the exhausted group finally reached the opposite side of the island, they found the basins filled only with more of the same rotting timber they had spent the past hour scrambling over.

Fuertes had struck out into the forest on his own. Anxious to bring back rare bird specimens for his paintings, he had looked forward to his first foray into the coastal wilderness. The young artist set out across the damp mossy logs, which provided the best means of moving through the underbrush. He immediately lost his balance on a slippery log and knocked his chest hard against the wood. More embarrassed than hurt, Fuertes quickly picked himself up before anyone saw him and proceeded again with more caution. Hummingbirds zoomed past his head, hovering inquisitively in the air before taking off again. The artist made quick mental notes of the colorings and markings on the birds so he could paint them from memory. The sound of a warbler enticed Fuertes farther into the woods, but he soon noticed an even stranger call. His curiosity aroused, he ventured in the direction of the unusual bird that could make so unusual a song. As the noises became increasingly bizarre, Fuertes began to suspect a prank. His eccentric companion Albert Fisher, he knew, enjoyed taking advantage of his youth, and often schemed to play tricks on him. The artist slipped quietly back down the path he had come, leaving Fisher to make his noises alone in the wilderness without satisfaction.

The rugged woods intimidated Fuertes' fellow artists Dellenbaugh and Gifford. Along with John Burroughs, the three took a small boat and paddled to the head of the inlet for a better view of a picturesque waterfall they had glimpsed when the *Elder* pulled into the dock. There the artists made quick sketches before the whistle signaled the ship's departure. William Dall hurried back to the ship equipped with hammers and mortars he had obtained from the Indians on the island and proudly displayed his new possessions to his companions on board. Many years of bargaining with the Alaskan natives had made Dall a skilled businessman in dealing with the Indians, who exchanged their goods only for the particular coins they thought most valuable. Albert Fisher also brought a souvenir on board—he had shot one of the region's enormous rav-

"Who Are We?" by E. S. Curtis. Harriman Expedition staff and family at an Indian village, Cape Fox, Alaska. *Front left, seated:* Wesley Coe, Trevor Kincaid, B. K. Emerson, Robert Ridgway (fifth, looking to right), William Ritter (black derby), Alton Saunders (bearded), unidentified man, Thomas Kearney, G. K. Gilbert, Henry Gannett, three Harriman daughters and their guests, Mrs. E. H. Harriman (on log), C. Hart Merriam (white Stetson). *Middle row, kneeling:* Charles Palache (holding arrow), A. K. Fisher. *Rear, standing:* Charles Keeler (third from left), William H. Brewer (fifth), Frederick V. Coville (tall), B. E. Fernow, Frederick S. Dellenbaugh, two unidentified women guests, George F. Nelson (chaplain), tall unidentified man, John Burroughs, Louis Agassiz Fuertes, Edward H. Harriman (behind wife), Lewis B. Morris (physician), William Trelease, George Bird Grinnell, Daniel G. Elliot, William H. Dall. John Muir was along but is not shown here, nor is the photographer E. S. Curtis. From: Alton A. Lindsey, "The Harriman Alaska Expedition of 1899, Including the Identities of Those in the Staff Picture," *BioScience,* June 1978.

ens, and Fuertes began sketching the bird as soon as he could wrest it from Fisher.

At New Metlakahtla, the expedition's next stop, Fuertes would discover a community filled with the strange, tame ravens. The ship's chaplain, Dr. George F. Nelson, who held a prominent position in New York as Bishop Porter's secretary, informed the scientists that the upcoming community was the site of a unique experiment in "civilizing the savages." On Annette Island, New Metlakahtla had become the settlement of a Scottish clergyman named William Duncan. The missionary

had brought a group of once fierce Indians from British Columbia to the island so that he could establish his own rather
unorthodox system of merging religion with capitalism. In
British Columbia, Chaplain Nelson explained, Duncan had
gotten into trouble with the Episcopal Church after he had
refused to use wine in the sacrament. The mere taste of alcohol, Duncan thought, would corrupt the childlike Indians.
When the church insisted that he conform to tradition, the
stubborn clergyman transported his entire flock of Indian con-
verts to Annette Island. At New Metlakahtla he assumed strict
guardianship over the natives, supervising them as they cleared
the land and erected, among other structures, a church and a
salmon cannery, which had become a thriving business. William Dall, who had named the island after his wife, Annette,
many years before Duncan's arrival, continued Nelson's discussion of the experiment in Christianity and capitalism at the
settlement. The scientists and artists looked forward to seeing
the unique compound firsthand.

They were not disappointed when, at nine o'clock on Sunday morning, the *Elder* reached Annette Island. There they
found the short, gray-bearded Duncan, whose size and age

Indians at New Metlakahtla.

belied the enormous power he wielded over his nearly one thousand charges. A large church with two tall towers over-shadowed the town, with its wide board sidewalks and mod-est, neat cabins. The careful, orderly construction of the place impressed the scientists as they surveyed the entire village, noting the schoolhouse, the town hall, and the sawmill, all with their promises of democracy, education, and economic industriousness. Duncan proudly invited the expeditioners to his church. The morning service brought the Indian commu-nity out in full, dressed in the fashionable hats and suits of white America. After a few perfunctory remarks from the church's pulpit by Dr. Nelson, Duncan took the stand and delivered an impassioned sermon in the Indian's native lan-guage, gesturing emphatically to impress his guests.

Duncan's role in organizing the Indian community amazed the Harriman expeditioners. Unable to understand the ser-mon, Keeler spent the time scrutinizing the Indians with a wide-eyed curiosity. Only thirty years earlier, he thought, Duncan's congregation had been wild cannibals; and now they sat together, miraculously enough, as "ladies and gentlemen." John Burroughs approved of Duncan's domain; the clergyman was, he thought, "really the father of his people. He stands to them not only for the gospel but for the civil law as well. He supervises their business enterprises and composes their fam-ily quarrels."

Such unrelenting "supervision" would lead to Duncan's de-mise several years later. Some fifteen years after the Harriman expeditioners had paid their congratulations to the patriarchal Duncan, the clergyman's charges rebelled against their dicta-torial leader and demanded authority over their work as well as a share in the cannery profits, over which Duncan had full control. The "quick, respectful curiosity" that Burroughs found in the Indians' faces became a reasoned, vehement demand for their economic and political rights by 1915.

But as they sat listening to Duncan's fiery oration in 1899, the Harriman expeditioners wholeheartedly approved of his experiment in taming the Indians with such seeming benevolence. John Burroughs thought it "one of the best object lessons to be found on the Coast." With an underlying sense of relief, the naturalist noted that the natives of Alaska seemed far more amenable to the demands of "civilization" than did most of the other original inhabitants of the United States. Smaller and lighter-skinned than the natives whom Burroughs had encountered previously, these Indians seemed more passive by contrast. The Aleuts, he thought, took "more kindly to our ways and customs and to our various manual industries." Certainly, the labor was available for the region's economic development if the natives could be endowed with a firm sense of Christian morality and the conscientious Protestant work ethic that "had built America." Harriman paid even closer attention to the matter than did his official historian, Burroughs. Speaking of the intelligence and industriousness of the Indians, Harriman observed that "if they could be taught to speak the English language, they could be largely used in the development of the territory."

In the age of budding imperialism, the vast reaches of the Arctic territory held a promise of economic profitability to the mind of an entrepreneur like Edward Harriman, and the natives who called the wilderness home could become central to industrial development. Even George Bird Grinnell, whose sympathetic contacts with the Indians spanned many years, heartily approved Duncan's success in making good workers of the Indians. He penned an accolade to the domineering clergyman: "It took many years for Mr. Duncan to change these Indians from the wild men that they were when he first met them, to the respectable and civilized people that they are now."

These positive assessments were characteristic of Victorian America's view of the Indian's place in society. No less an authority than John Wesley Powell, founder of the Bureau of American Ethnology, believed that Indians should be taught the skills of farming and mechanical trades as well as the Protestant work ethic. This had likewise been the thrust of all the Indian schools, such as that at Carlisle, Pennsylvania, and the philosophy had been incorporated into federal law with the passage of the Dawes Indian Severalty Act of 1887. In the Gilded Age, "Americanization" of the Indian was a prime objective. This meant training in English, inculcation of the work ethic, and Protestantization. There was little thought, among Bureau of Ethnology scholars, of the need for maintaining tribal integrity and tribal customs; they were more interested in preserving the history of what they deemed to be a way of life that would inevitably vanish before the onset of "higher" stages of civilization. Thus the expeditioners, in their view of New Metlakahtla, were right in line with the social Darwinist thinking that characterized Victorian America. In their admiration for Duncan's reform experiment, they overlooked entirely the damage done to the fragile Indian culture. They saw only "savages" converted to the higher calling of productive civilization.

John Muir, however, withheld judgment on the prospering little community over which Duncan reigned. The son of a stern Scottish clergyman, Muir frequently entertained his company with tales of his harsh upbringing at the hands of his father, and he saw, perhaps, a little of his own dictatorial father in Duncan. Muir escaped after the church service into the island's lush forests with his friend John Burroughs, and the two naturalists took a leisurely pleasure in counting the varieties of birds, flowers, and trees they sighted. The island's tame ravens fascinated Burroughs, and from their perches along the roofs of the Indians' carefully groomed homes the noisy birds raucously scolded the town's guests. Totem poles erected by the island's inhabitants years earlier cast long shadows over the town, providing an ever-present reminder of the Indians' original beliefs before men like Duncan migrated to the region.

The presence of a group of white people on the island was a rare event. Since Duncan feared the corrupting influence of gold miners, he had made a firm policy of arresting on the spot any white person who landed on Annette Island. The prospectors, he surmised, with their lust for gold and liquor, would undermine his efforts at taming the Indians. Determined to protect his haven from the greed of the miners, Duncan fought against the legal efforts in far-off Washington, D.C., to open his isolated community to prospecting. The clergyman ran a rigid colony with its own police force, based on an old agreement with the United States government that his settlement on Annette Island would be legally protected from harassment. George Bird Grinnell shuddered at the thought of what would happen to Duncan's Indians if gold miners were allowed to invade Annette Island.

Totally unconcerned with Duncan's religious efforts, one expeditioner ignored the church service to take a long walk in the woods by himself. Albert Fisher, in fact, became so enraptured with the birds and animals on Annette Island that he was still lost in meditation in the forests when the *Elder* pulled away from the dock. Fortunately he was soon missed, and the ship returned to fetch the ornithologist; Fisher casually emerged from the woods hearing the familiar whistles of the *Elder*. Despite the cries of his colleagues to hurry on board, the ornery scientist refused to quicken his step, and slowly made his way down the shore to the steamer, where his shipmates concealed their annoyance with teasing remarks. Fisher, they claimed, would never die of heart disease with his rate of physical exertion.

With the roaming ornithologist safely back on board, the Executive Committee turned its attention to devising a system by which the scientists could officially "check out" when, like Fisher, they struck out on their own. Dellenbaugh carefully painted the names of all the expedition members on a large board, with pegs following each name that were to be moved from spaces labeled "aboard" to "ashore" when they ventured alone into the wilderness.

Dellenbaugh finished his task in time to attend the Sunday evening service of hymn singing, but the gathering soon dispersed to watch a glorious Alaskan sunset. Dall, ecstatic at the sight of the magnificent mountain peaks radiant with the rays of the sinking sun, declared that here, at last, was true Alaskan scenery. He happily informed his colleagues that the Gulf of Georgia sunset they had admired earlier was merely British Columbia scenery. Brewer caught his breath as he tiptoed out of the religious service into the brilliant twilight "alpenglow." The jagged, snow-covered mountains along Clarence Strait seemed to vibrate with golden light, and Merriam, who often seemed to feel responsible for the scientists' enjoyment of the expedition, watched the group's pleasure with satisfaction. But a joyous Burroughs, elated at the gorgeous sunset, could not resist a jab at John Muir, who drifted onto the deck after the sun had already set. He gleefully chided his friend, saying that "you ought to have been out here fifteen minutes ago instead of singing hymns in the cabin." Muir, never to be outdone, quickly shot back at Burroughs in his thick Scottish accent, "Aye, and you, Burroughs, ought to have been up here three years ago instead of slumbering down there on the Hudson!"

The sunset inspired Fernow to ripple through one of Beethoven's sonatas on the ship's piano. Swain Gifford and Charley Keeler sat enthralled by their companion's talent as the music wafted throughout the *Elder*. After the impromptu concert Keeler met Charles Palache on deck and recognized the signs of homesickness in the young scientist's face. Palache confided that his bride-to-be waited back on the East Coast so they could be married as soon as he returned. Keeler produced photographs of his wife and young daughter to show his new acquaintance, while Palache talked of his wedding plans.

Keeler was already tiring of the immense effort involved in keeping up with the expedition's scientists and their constant talk. Emerson reveled in telling his companions tales of his world travels, while Brewer, Keeler noted, was "always trying to waylay somebody to tell a story." Albert Fisher's pranks and Merriam's affinity for ribald, salty stories added to Keeler's sense of confusion. He was not sure of his own place among such men; Burroughs, it seemed, was appointed official historian and Merriam's request that Keeler "contribute something to the literary part" of the voyage was now altogether unclear. In a rare moment of privacy, Keeler sat on the trunk in his stateroom feeling rather dejected that his position remained vague. But he was not alone with such thoughts. Two doors down from Keeler's room, another expedition member often sought refuge from the crowd of exuberant scientists. To John Burroughs, his companions seemed "fearfully and wonderfully learned—all specialists," while he felt "the most ignorant and the most untravelled man among them, and the most silent." At such times, the looming mountain ranges that marked the inside passage to Wrangell seemed only to remind the lonely Burroughs of the peaceful rolling hills that he had left behind on the Hudson River.

Early the following morning, the ship steamed into the harbor at Wrangell. With only a few hours to investigate the new terrain, the scientists set off in small parties to carry out their tasks as quickly as possible. The botanist Alton Saunders, rose at three and beneath a clear, cold sky probed along the shoreline at the hour of low tide. "Seaweed Saunders," as John Muir teasingly called him, picked his way among the Indians' large dugout canoes—carved from single trees—that lined the shore, collecting algae and seaweed. He had already been exploring the shoreline for a couple of hours when he saw the ornithologists marching off in a small group into the forest, bearing their guns to fetch specimens of the region's bird life.

After wolfing down a breakfast of strong coffee and crackers, Merriam trekked off alone into the bogs south of the small community and found them covered with dwarfed trees and bushes. He was pulling down vegetation for his collection of plants when an injured bird fluttered past. Never without his gun, Merriam with a single shot bagged the stellar jay that Fuertes had managed only to wound. The flowers and birds enchanted John Burroughs, and he winced as the sounds of shots echoed across the town. Shivering beneath an outsized overcoat he had borrowed to keep out the penetrating cold, Burroughs reveled in the songs of the town's sparrows and swallows, which sounded, he found, remarkably similar to those of the birds that serenaded him as he worked in his vineyards at home. For his companion Muir, however, the Alaskan forests *were* home, and he ambled slowly along the outskirts of the woods, carefully measuring the grand trees for his personal records.

The town of Wrangell itself attracted Frederick Dellenbaugh, and he busily snapped photographs of the decaying totem poles that stood in front of many of the small wooden houses. Built on the Stickeen River, the forlorn town had once been thought to be a gateway to the gold fields of the Yukon, but the community had suffered the fickle ways of the mad quest for the yellow metal. Property values, which had rocketed a year earlier as the gold prospectors surged through town, had fallen off sharply when a new route to the North was discovered. A row of once-active stern-wheel steamers testified silently to the town's decline, as they lay "peacefully decaying on their skids." Desolate little Indian homes stood near the shoreline on stilts that precariously straddled the high tide. The natives' wolflike dogs slunk through the town, howling mournfully at the strange men who invaded the marshes and woods with their instruments and notepads. Deserted restaurants, saloons, and hotels gave evidence of the wave of prosperity that had washed through Wrangell and left it barren. A brewery and sawmill were the town's only businesses. Fascinated by the community's rapid decline, Dellenbaugh mused that, in time, Wrangell would probably enjoy economic prosperity again. Charles Keeler was not so optimistic. Still despondent over his uncertain role in the expedition, he wandered down the town's streets, which consisted of boards raised on stilts to protect them from high waters. Wrangell, he sniffed, was a "dirty, miserable" town, its ditches filled with old shoes and tin cans discarded by the miners who hurried through on their way to the gold veins. Even the attractive totem poles, he noticed disdainfully, stood "close beside the most hideous monstrosities of painted modern houses, all dirty and dilapidated—the houses of the Indians." Keeler gladly boarded the *Elder* when its whistle sounded sharply at eight o'clock, signaling the ship's departure.

Chief Snake's House, Wrangell, by E. S. Curtis.

But Keeler's spirits lifted in the late morning as the ship cruised slowly through Wrangell Narrows, lined with spectacular snow-covered mountain ranges and magnificent glaciers descending down into the water. The cold, brisk air and clear sky exhilarated the passengers on the *Elder,* and even the dignified Harriman pranced along the ship's deck with his two young sons, eight-year-old Averell and two-year-old Roland. Burroughs glanced up from his writing just as his host raced past the door of his stateroom. Harriman's unpretentious, direct manner impressed the naturalist, and Burroughs took the occasion to scrawl in his letter that the railroad tycoon seemed "very democratic and puts on no airs."

On the deck, several scientists amused themselves with an impromptu target practice, aiming their guns at the numerous ducks, eagles, and gulls flying alongside the ship. Bright explosions of gunshot from the ship's deck ripped across the still clear waters as the *Elder* glided past majestic glaciers that dwarfed the noisily intruding ship. The Devil's Thumb, a steep, forbidding mountain with a rugged shaft of granite rising one thousand six hundred feet beyond its peak, loomed ahead. John Muir pointed out the inaccessible landmark to his companions as they steamed past on the way up to Juneau. For Muir, the day was incomparable, offering clear views of the mountains unhindered by the region's usual fog and rain. The ship passed slowly along the coast, and Muir likened the experience to "turning over the leaves of a grand picture book." He was pleased when his fellow passengers laid down their guns to watch in awe as the glaciers royally unfolded before them and icebergs drifted slowly past the hull of the ship.

With Patterson Glacier appearing like a frozen river in the valley of a mountain to the right of the *Elder*, Muir explained that the largest glaciers lay still ahead. In fact, he boasted, the upcoming Muir Glacier, which he had discovered in his Alaskan explorations years earlier, held more ice per square mile than all of Switzerland's glaciers combined. But even the comparatively small Patterson Glacier seemed to stretch endlessly into the icy wilderness, and Muir enjoyed the opportunity to tell the scientists of the twenty-mile-long hike he had once taken to trace the head of the Patterson. John Burroughs, responding to Muir's self-professed expertise on the glaciers, labeled Patterson Glacier as "the smallest lamb of the flock of Muir's Mountain Sheep." Indeed, Muir so often volunteered his knowledge of the glaciers that the company soon wearied of his spontaneous lectures, particularly when they found their own opinions quickly discounted by the argumentative Muir.

Burroughs turned from Muir's storytelling to study the birds visible on the coastline. Squinting at the horizon, he counted seven eagles regally perched in a row on the shore's rocks, looking, he wrote, like "Indian chiefs" whose piercing eyes haughtily surveyed the ship. Whales splashed playfully near the ship, blowing tall fountains of ocean water from their spouts while the biologists attempted to estimate their size. The Indians' ritualistic whale-hunting expeditions had not diminished the numbers of the great creatures, which supplied the basic elements of food and fuel for the natives' existence.

Early in the afternoon, the steamer anchored in Farragut Bay and two parties of scientists lowered boats to the water. William Brewer helped his companions place their traps on the floor of the small boats that would carry them ashore for a two-hour foray into the woods, where they hoped to catch specimens of the region's small-animal life. From his vantage point on the ship's deck, Muir had recognized the woods along the shoreline as a site which he had explored some twenty years before, and he stepped into one of the scientists' boats for the ride to shore. Muir was eager to ramble again through the familiar wilderness and assess the changes it had undergone in the intervening years. Aided by Dr. Morris, Merriam set traps and then set about exploring the woods. Deer and bear trails laced through the trees. At one spot, the two men discovered a clump of wolf hair alongside the carcass of a deer, and, farther on, Merriam found antlers and bones, which he gathered to add to his assortment of specimens. While the two men picked through the remains, an eagle descended from its nest high in the top of a Sitka spruce, winging its way over their heads as if to investigate for itself this strange breed of human animal. Merriam watched the huge bird jealously, coveting the fine specimen it would have made.

Grinnell wandered off by himself, picking his way through the fallen timber to explore the deeper recesses of the woods. At the sound of the ship's whistle he hurried to retrace his long hike through the dense forest, but as the minutes passed, he worried that he might have ventured too far into the woods to make it back to the shore on time. When he finally stumbled out onto shore, he saw his companions paddling rapidly off toward the *Elder* in the distance. The ethnologist let out a loud yell and to his relief one of the boats came hurrying back to rescue him.

It had been an unlucky trip to shore for the biologists, who disgustedly tossed their empty traps back on the *Elder*. Two hours had not been sufficient time to catch the animals they needed for study, but there would be another opportunity later in the afternoon, when the ship planned a stop at the Indian village of Taku. For Harriman, the news of numerous bear trails on shore seemed to assure good hunting, and he was anxious to explore on his own for evidence of the region's famous bears. At some point along the coast, Harriman planned to venture inland in pursuit of one of the legendary beasts that stalked the wilderness. A rich bearskin would provide the ultimate souvenir of his luxurious vacation, and Merriam's and Morris's word of bear tracks on shore traveled quickly among the expedition's hunting enthusiasts.

The two-hour break at Farragut Bay had been productive for Dellenbaugh. While his colleagues explored the woods, he remained on board to paint the high mountains rimming the bay. Pleased that he had had time to complete the picture before the scientists returned, Dellenbaugh was busily putting the finishing touches on the work when the ship's steward passed by. Amazed at the artist's quick work, he exclaimed, "Say, you're quite a flyer." Dellenbaugh's experience on western exploring expeditions had trained him at an early age to work rapidly, pacing his schedule to match that of the scientists he accompanied. The *Elder* had drifted aimlessly while the scientific parties worked, and Charles Keeler, who found writing difficult amid the usual activity on board, had also put the quiet time to good use. Before the scientists clambered back on deck, he tucked away his journal and signed a letter to his wife.

With the expedition scheduled to tour one of the world's largest mining operations, the Treadwell Mine near Juneau, the following day, Walter Devereux lectured his colleagues on the mine. Its innovative techniques in excavating vast quan-

tities of low-grade gold-bearing quartz had made the operation world-famous. Standing at the podium that Harriman's
Committee on Lectures had erected on the hurricane deck,
Devereux explained how recent technological developments in
mining machinery had turned the region's previously unprof-
itable quartz into a valuable resource. Though a ton of quartz
brought only three to four dollars, the new machinery and
efficient methods extracted such enormous amounts of the gold-
bearing ore that the operation had become extremely lucrative
for the mining company.

After Devereux's talk, the ship anchored at Taku Village,
and the biologists lowered a boat for a second try at trapping
the rodents they sought. Harriman joined the scientists as they
crowded into a single small boat with a hundred of their traps
scattered around them. Working rapidly in the dying light of
the long Alaskan day, Merriam, Fisher, and Palache set out
the traps and then leisurely explored the old village with their
companions. In the growing darkness, the men found an almost deserted town, inhabited by a grizzled old trader, a few
Indians, and hundreds of ducks. Muir, always acting as the
authority, explained that the Indian tribe that once populated
the community was now nearly extinct. Their numbers had
slowly dwindled through the years as white settlers and explorers brought diseases with them. He remembered sadly that
twenty years earlier, one hundred Indians had lived at Taku,
but by 1899 ruins of small cabins marked the last traces of
their existence. Moved by the poignant sight, Edward Curtis
silently began photographing the village remains in the last
light, following Harriman's advice to make good use of Alaska's
long periods of sunlight. Curtis' companions, meanwhile, ventured inside a house that Muir had pointed out as one of Taku's
many "houses of the dead." Here, where the Indians deposited the remains of their dead after cremation, the men found
trunks and boxes filled with charred human bones. Muir informed his company that all but the tribal shamans were cre-
mated; the shamans, Merriam irreverently noted were "boxed
up raw."

The small party left the desolate scene at midnight as darkness finally settled. They silently paddled across the still bay back to the ship, scheduled to depart for Juneau the next morning. After only a brief rest, Fisher drowsily knocked on Merriam's door at four a.m. so they could collect their equipment and gather their trapped animals on shore. For the two sleepy scientists it was a needless trip. Stumbling along the shore, they discovered trap after trap empty. Much to the amusement of their colleagues, Fisher and Merriam returned to the *Elder* bearing a meager catch of five mice and one toad.

Several hours later the small town of Juneau appeared in the distance, picturesquely tucked at the foot of a snow-capped mountain. Juneau's two thousand residents, most of whom were Indians, catered to the wealthy tourists who filed through town during the summer months. After mailing their letters home, most of the Harriman expeditioners explored Juneau's numerous souvenir stores. Dellenbaugh bought a small replica of a dugout canoe carved by Indians to decorate tourists' homes in the States. John Burroughs found time to introduce himself to two local men who, to the nature author's immense delight, expressed their admiration for his books. Even in the remote regions of the Arctic, Burroughs thought smugly, his writing found fans.

Across the Juneau on Douglas Island, the Treadwell Mine thundered and roared with the new equipment that Devereux had praised so highly. Even in 1899, discharge from the great mine already polluted the crisp Alaskan air. A single blast from Douglas Island ripped across the water so loudly that Dellenbaugh, sitting in his stateroom as the ship moved toward the island, thought someone had fired the small cannon on the hurricane deck. Once they arrived on the island, deafening sounds of heavy machinery as it crushed tons of quartz overwhelmed the expeditioners. They peered down into the deep shaft of the mine, where a gaping hole several hundred feet wide spread before them. With the sound that Dellenbaugh had thought to be a cannon shot, constant explosions shook the earth. Covering their ears against the noise, the party observed the miners working below in the great pit, dwarfed by the steep walls towering far over their heads. The working conditions stunned Dellenbaugh. The workers looked like the pygmies of Africa, he thought, toiling for their meager wages of two or three dollars a day. A company manager, oblivious to labor or ecological problems, explained that all the neighboring mountains contained such potentially profitable gold ore, but the dense forests were an obstacle to the industry.

The terrible roar became unbearable to the guests; for Burroughs, Niagara Falls was only "a soft hum beside it." Dall, not eager to lose his hearing, quickly escaped the noise and set off to explore the island. The forests that once covered Douglas Island had been reduced to unsightly stumps so that the mining company could expand its operations. Dall made his way through the cleared areas in search of fossils in the terraces that rose behind the mine. The old explorer soon rejoined the party, bearing several fossilized skulls he had unearthed not far from the mine. With their hearing numbed by the thundering of the mine, the group gladly boarded the ship for its next stop at the gold-rush town of Skagway. Five scientists remained behind on the island to study the area more carefully than the ship's schedule allowed. Palache, Ritter, Saunders, Kearney, and Kincaid took a steam launch from the *Elder* and headed off to make camp on shore for several days.

As the *Elder* puffed away, Harriman noticed a scraggly dog scampering about the deck. A passing crew member informed the tycoon that the dog had followed another sailor on board at Juneau. Harriman frowned and commanded, "As long as this dog remains on board, he is our guest." With those words, he marched off to find the sailor who had befriended the animal and order him to feed his pet well until it could be deposited again at Juneau. No detail escaped Harriman's scrutiny, and, as the entomologist Kincaid had already noted, he "was of the type that issues orders and expects them to be obeyed." The dog ate well during his brief voyage on the *Elder*.

Indeed, in an age of sentimentality, it almost seemed as if dogs fared better than people on the Alaskan frontier. By this time the expeditioners had had ample views of the juggernaut of "civilization" invading the pristine wilderness: they had seen the "guided democracy" of New Metlakahtla, assembly-line canneries served by cheap coolie labor, and the rotting ghost town of Wrangell together with its Indian village counterpart, and they had "experienced" the technological marvels of the Treadwell Mine, grinding away at forest and mountain alike. While they admired Alaska's natural beauty and were curious about its natives, they seemed to regard the advance of exploitative civilization as a necessary inevitability. As scientists they believed in progress, but being of the "better sort," *they* could appreciate nature, too, and they could not help but feel good about their sense of compassion. By all means save the stray dogs.

CHAPTER FOUR

THE STUFF OF LEGENDS: SKAGWAY, THE GOLD RUSH, AND DEAD HORSE TRAIL

At the close of the nineteenth century Alaska's wilderness not only lured elite tourists, professional scientists, and U.S. Senators seeking a fashionable vacation spot, but its vast frontiers also held the promise of quick wealth for the gold miners who flocked there after 1897 by whatever means they could afford. The gold rush had left Alaskan land trampled by the tired feet of hopeful men and women who had sought fortunes like those that Edward Harriman and others had already accumulated. In this industrial era riches represented the American dream, and men such as Harriman, who panned for their gold in corporate boardrooms, and the rare gold miner who actually struck paydirt symbolized the possibilities of success for every American.

At the end of the Lynn Canal, the next destination of the *George W. Elder*, stood the lively town of Skagway, one of several boom towns that had sprung up two years earlier to serve the transient population of miners. After spending an afternoon leisurely surveying the grand sights along the canal, the Harriman Expedition headed directly for the gold-rush coun-

try. During their tour of the Treadwell Mine, the expeditioners had witnessed firsthand the massive corporate attempts to
reach the gold ore beneath the land's dense timberlands. At
Skagway they would walk among the lonely, rugged miners
who sought gold on their own terms, often on tragic treks
through the remote and dangerous regions that offered dreams
of fortune.

John Muir half-dreaded the stop at Skagway. On his last
visit to the town two years earlier, he had found the gold miners like "a nest of ants taken into a strange country and stirred
up by a stick." Alaska, he worried, might be ruined by the
unbridled greed for gold.

In Skagway, there was not one stick but two—made of
steel—to excite the miners. The tracks of the newly completed
White Pass Railroad glittered as they ran to the summit of
White Pass. By foot, it was a long, back-breaking hike to the
pass for miners lugging camping goods uphill through the
snow. But the railroad now offered a safe, quick trip through
the Alaskan interior along a twenty-one-mile portion of the
gold route. The construction of the line, however, had hardly
been safe for the workers who actually built it. The path of
the railroad had been bloodied by many accidents during the
wintery months of 1898–1899. Falling rocks and explosions
had killed thirty-two men—at least these were the deaths admitted by the White Pass managers.

Edward Harriman had already negotiated with White Pass
Railroad officials to ensure that his guests could ride the length
of the line. He was extremely curious about this daring little
railroad, for the success of the engineering feat promised, perhaps, even more rail ventures into the Alaskan interior. The
foresighted builders of the White Pass line had weathered ini-
tial ridicule of their project when critics assailed their proposed work as rash and foolish, but now the officials heard
only praise as they raked in profits on their gamble. They
already planned extensions of the line farther into the rough
territory; certainly, if the gold rush continued, the demand for
railroads to transport gold miners would rise. The question

remained, however, of the viability of railroads across the treacherous Arctic terrain, and Harriman was eager to assess the potential of similar lines.

All day the *Elder* steamed up the straight fjord of the Lynn Canal, heading for Skagway. The town, built at the mouth of the Skagway River, provided gold miners a jumping-off point for the Klondike. Skagway stood at the entrance to the White Pass, a thin path over the mountains that led to the gold-rich Yukon River. In the two years since the first streams of miners had begun pouring into Skagway, the town had endured wave after wave of gold-seeking fanatics. The sudden rush on the wilderness town left it a prime spot for violence and chaos, as newcomers frantically tried to transport their possessions through the White Pass, and the "old-timers" who had spent several months in the area exploited the newest arrivals. Men like the notorious racketeer "Soapy" Smith, who had ruled the town with a dictatorial hand until he was shot down in the streets only the year before Harriman arrived, had lent Skagway a reputation for lawlessness. Word of the colorful community quickly filtered down to the States, and newspapers hastily tried to cover the events. One daily in San Francisco had naïvely approached John Muir to hire him as a correspondent in the Yukon territory. Muir, appalled at the very thought of a gold rush through his beloved wilderness, haughtily refused to participate in the ensuing publicity. Such a reporter's job, he thought, was a "sordid mission" in a land that could easily be spoiled by mad crowds of miners.

One young writer had already taken up the journalistic challenge, however. Lured by the "call of the wild," Jack London had ventured to the Yukon in 1897 and returned to California a year later, richer in experience than in gold. His encounter with the harsh world of the Alaskan gold country had been far removed from that of the Harriman Expedition, and London's accounts of the gold miners' desperate struggles for survival immortalized that crucial Alaskan era.

In countless stories and in *The Call of the Wild*, London depicted the degraded Indians and defeated sourdoughs against a backdrop of brutal social Darwinism in a way that the Harriman expeditioners glimpsed only on occasion. They were largely optimistic about man's possible redemption and the inevitability of progress and refinement. London, who had lived with the gold seekers, dreamed their tragic dreams, starved and suffered scurvy with them, and fought for life alongside them, was not so optimistic. He saw the Klondikers reverting to savage "tooth and claw" primitivism in a struggle for survival that reduced men to mere brutes considerably less noble than their sled dogs. To him the development of *this* side of human nature in the wilderness was inevitable. Skagway would provide the expedition's introduction to this lurid world, from which London had only recently fled.

When the *Elder* floated into the bay at Skagway around ten in the morning, a motley crowd of excited townspeople swarmed along the shore, confirming the town's colorful legends even before the ship docked. They waved madly to the scientists gathered on deck. Here at last, thought the Skagway residents, was the mail boat *Rosalie* bringing them long-awaited letters from home. And, from the ship, the baffled scientists thought all of Skagway had turned out to welcome their prestigious group. As soon as the ship touched the long pier, more than a hundred men and boys scrambled recklessly aboard and raced wildly across the decks of Harriman's ship while the scientists watched, amazed at the sudden chaos. In stark contrast to the repressive order they witnessed at New Metlakahtla, the lusty, unruly crowd spilled across the decorous decks of the *Elder* until the ship's crew sternly ordered the intruders off. John Burroughs' eyes bulged, possibly not with innocence, at the sight of women in billowy "bicycle suits," who "gazed intently at the strangers." The frontier, it seemed to the naturalist, had spawned a bold new breed of woman.

Once the excitement had ebbed, the expeditioners cautiously disembarked to explore the lively town before dark. Built on the sandy flats at the base of the steep White Pass ridge, the town's business district boasted flimsy hotels, saloons, and restaurants that haphazardly lined rutted dirt roads.

Hotel "runners," hoping for a flourishing night's business with the well-dressed new guests, called out the names of their various lodging quarters. Dellenbaugh sighed with relief that the *Elder* provided a "floating hotel" for their stay.

Though Muir had steeled himself for the sight of the "ants' nest" waiting at Skagway, he looked forward to the possibility of locating an old friend, Hall Young. Young was a minister whom Muir had last seen in Skagway in 1897, and the naturalist hoped that his friend's religious work had not taken him elsewhere. As luck would have it, Young had remained in Skagway, and the two friends enjoyed a reunion during Muir's brief stay in town. Muir and Young had first met in Wrangell in 1879 and immediately struck up a friendship that the intervening years had not diminished. The older naturalist, Young claimed, had revealed to him the spiritual harmonies of nature, and, years later, Young still felt so moved by his friend that he wrote, "I sat at his feet; and at the feet of his spirit I still sit." Muir had unfolded the "secrets of his 'mountains of God'" to the minister. Young had accompanied Muir on several of his many excursions into the remotest regions of Alaska, including his famous adventure on Muir Glacier with the legendary dog Stikeen, of whom Muir never tired of talking. The minister's rapport with the natives of the area had taught Muir the importance of friendship with the Indians, whose knowledge of the rugged terrain had played a central role in Muir's discovery of Glacier Bay with Young in 1879. In 1897, when Young began his work among the first gold miners at Skagway, Muir had warned his friend that "instead of the music of the wind among the sprucetops and the tinkling of the waterfalls, your ears will be filled with the oaths and groans of these poor, deluded, self-burdened men." In such an environment, the naturalist suggested, it would be important to remain "close to Nature's heart" and to keep "your spirit clean from the earth stains of this sordid, gold-seeking crowd in God's pure air." Only then, Muir claimed, could Young "bring to these men something better than gold." Young had worked among the miners, often remembering Muir's advice; but when Muir suggested that Young join the expedition for a respite from his work, the minister reluctantly refused the offer. The mining camps of the interior needed him, he explained, and soon he and several other ministers would be bound down the rough Yukon for the remote outposts of the gold territory.

While Muir visited with his old friend, Gannett, Fernow, and Dellenbaugh surveyed the town in the last of the day's dying light. The great tree stumps jutting up throughout the town and the long rows of new cabins made Skagway "one of the best illustrations of a new frontier town" that Dellenbaugh had ever seen. In only a year, entire spruce and hemlock forests had been cleared to build a way station for the hordes of northward-bound travelers. Rows of new houses stretched out from the periphery of town, following the shining new tracks of the White Pass railway that ran past them on its way up to the pass. In the morning, the Harriman party would ride the line as guests of the railroad, and most of the expeditioners turned in early, assured that tomorrow would afford a closer look at the legendary Skagway.

A drizzling, rainy morning greeted the expedition members when they rose on June 7 to eat breakfast before embarking on the three-foot-gauge railroad that would lift them to White Pass. As the newly completed railroad sped along the Skagway River, the haze lifted enough for the passengers to see what Fuertes called the town's "suburbs" of log cabins. Slowly, the railcars began to climb the steep banks of the mountain range behind the gold-mining town, and soon they had reached the infamous Porcupine Hill. The site of the grisly Dead Horse Trail, Porcupine Hill had been jammed with gold miners during the deadly winter of 1897–1898, and the lines of prospectors and packhorses had grown so thick that the stampeders spent hours at a complete standstill. Pushed beyond their endurance, thousands of horses died on the trail, and their owners continued without them. Animal carcasses, preserved by the snow and ice, still littered the mountain. The sight of two horses, their stiff legs sticking straight up into air, sent a chill down Dellenbaugh's back. The artist found it difficult to believe that such massive tragedies had occurred in the recent past. Many gold prospectors, he knew, had fallen victim to spinal meningitis and many more perished from exposure and exertion. Skagway residents told stories of the "early days of the gold rush," which, amazingly, referred to an era within the past three years. The long harsh winters and traumatic struggles for survival made the time stretch endlessly for the miners, and to some, it seemed that they had already spent a lifetime in their two or three years in the Yukon.

But the Harriman guests traveled the rough terrain of the morbid Dead Horse Trail in comparatively luxurious style. For twenty-one miles, the expeditioners chugged up the grade of the White Pass Ridge, a trip that offered scenes of stark rock and gorges that looked both "terrible and sublime" to Burroughs. Granite blasting done by the railroad engineers seemed to Burroughs to reveal the very "ribs of the earth"—a fitting scene for the grisly struggles of both gold miners and railworkers.

Merriam startled his company when he suddenly leaped from his seat and pointed out the window to a party of three men. "There are some of my boys!" he cried, waving to the small band. The men, he explained, were colleagues of his from the U.S. Biological Survey. The White Pass engineer cautiously backed the small train up to meet the group of Washington scientists. They were hiking up the mountain in gold-miner style, offering a stark contrast to Harriman's grand expedition. Government funding allowed for no such amenities as railroad passes. Headed for the Yukon and Dawson City, the three scientists were on a mission to study the animal life of the Alaskan interior. Wilfred Osgood, who later became Curator of Zoology at Chicago's Field Museum, where Daniel Elliot had his headquarters, greeted his old friend Merriam like a fraternity brother. The young Fuertes noted with pleasure that the distinguished zoologist seemed equally glad to see him, after their few encounters on the East Coast at scientific conventions. Merriam's friends boarded the train for the ride to the pass, amid lively chatter about their findings. Months later, however, Osgood's mission would meet with misfortune when a boat accident on the turbulent Yukon River caused them to lose all their specimens.

Once at the top of the White Pass, the train rumbled to a stop. The railroad's lines had just been completed all the way to Lake Bennett in the interior, but it was not yet in operation. Lake Bennett was then reachable only by a dangerous hike through thirty miles of mountainous terrain. The lake provided a gateway to the Yukon River, and from that point the gold miners could travel down the icy waters of the Yukon on steamboats to Dawson City. Once the White Pass railroad was open all the way to Lake Bennett, the route to the gold fields of the north would be far safer than it had been for the men and women who suffered the long journey on foot up Dead Horse Trail. Dall remarked that a traveler soon could check his bags all the way from Seattle to Dawson City. Ironically, by the time the rail opened its route, the gold rush had changed its course: Nome, at the far northern coast of Alaska was soon to draw a new stampede, away from Dawson City and Skagway.

At White Pass, some of the scientists quickly dispersed, striking off on their own through the scrubby spruce trees and granite rocks. Fuertes immediately spotted a bird which he took to be a golden-crowned sparrow and sighted it through his rifle, but the young artist was too close to his prey and his quick shot destroyed the specimen. The loss sickened him, but soon he bagged another sparrow, buoying his spirits. Several of his colleagues, however, surveyed the summit of White Pass with less enthusiasm. From the peak of the ridge one could see a small shanty village—White Pass City—composed of a dozen or so canvas huts huddled forlornly in the valley where the wagon trail ended and the rough mountains ahead forced the miners to move onward by packhorse. A cold, damp wind blew against the expeditioners' faces as they examined the desolate summit. "Godforsaken," muttered Dellenbaugh to himself, as he shivered against the cold wind. Burroughs

Flags at the United States/Canadian Border at White Pass.

conjured up all his best Ruskinian images to describe the trip up to the summit. He recalled, "The terrible and the sublime were on every hand. It was as appalling to look up as to look down; chaos and death below us, impending avalanches of hanging rocks above us. How elemental and cataclysmal it all looked!" He added for emphasis, "I felt as if I were seeing for the first time the real granite ribs of the earth. Here were the primal rocks . . . that held the planet together."

Another small collection of tents was perched on the summit, and the ragged flags of the British Empire and the United

States waved together in the breeze, signifying the boundary between the Canadian and American territories. Despite the image of the flags waving peacefully side by side, the site had been the object of a vicious dispute over the precise national boundary. Only the tattered remains of the American flag hung from its pole, since the United States refused to replace it, claiming that its territory actually extended twelve miles farther east. The governments haggled endlessly over the additional few miles of land. The Canadian police collected duties from the American miners on land the United States claimed as its own. Dellenbaugh bristled at the audacity of the Canadian government, which seemed, he charged, "bent on getting as much of our territory as they can."

Entering one of the little ramshackle, snow-covered canvas tents at lunchtime, the Harriman guests were surprised to find a table laden with a home-cooked feast. The railroad officials of the White Pass had made sure that Edward Harriman and his party would not leave their modest little company unimpressed with their hospitality. After they had finished the last of the pies and cakes offered them, Harriman called over to Edward Curtis to take a picture of the entire group as they sat at the long table. Try as he might to meet his host's request, the photographer could not make the dim light inside the cabin adequate for his equipment. The party then embarked once again on the railcars and the train chugged back down the steep pass, but a small group of rather important expedition members had been accidentally left behind. When one scientist noticed with a jolt that Harriman was not among the company in the train, the cars made a quick return to pick him up, along with his brother-in-law William Averell, who had wandered off at the pass to explore the area.

With everyone now safely aboard, the journey downward passed quickly. Talk of the gold rush prevailed, and tales of gold-mining episodes enlivened the ride. George Bird Grinnell, always on the lookout for a good anecdote, passed along a story that demonstrated the greed of the strange men who left their homes in the States to search for gold. So concerned with possessions were the miners, he said, that a man might begin his journey with one coffee pot—certainly all he needed, Grinnell interjected—but would continue to collect discarded pots until he burdened himself with as many as one hundred coffee pots, all stuffed into his packs or dangling from his belt. Here, remarked the story-teller, was an excellent "illustration of the needless accumulation of wealth." The similarity to Harriman's accumulation of railroads apparently escaped Grinnell.

In an attempt to best Grinnell's coffee-pot story, one scientist gave an account of an exchange between a Juneau gold miner and a visitor. Asked by the visitor if the rain ever stopped in Alaska, the miner replied, "How the devil do *I* know? I've only been here two years!" As the jovial party bantered, Burroughs again retreated into his own sentimental thoughts about the gold miners. On occasion, the train passed solitary sourdoughs on foot and the naturalist sympathetically noticed their heavy packs and the wistful expressions they wore when they glanced up at the train.

Several expeditioners debated the potential riches of the Klondike. Miners had excavated thirteen million dollars in gold from the region in 1898 and experts estimated that even more would pour from the streams and hills of Alaska in 1899. But the estimates, one scientist pointed out, were probably low, since many miners attempted to dodge the high royalties demanded by the Canadian government by slipping their gold out of the region without reporting its worth.

It was not the gold miner, however, who ensured Skagway a prosperous future, thought Dall. Tourists would bring the bustling town economic security. Nestled in his comfortable seat on the railcar, the solemn, taciturn Dall puffed on his ever-present pipe and reflected, without a trace of irony, that Skagway was indeed a model "town of the future," since the new railroad would lure tourists. After the expedition ended, the Alaska expert remembered the trip to the summit nostalgically, and prophesied that "it will not be long before a ride to the summit will form a part of every well-conducted tourist trip."

John Muir had sympathy for neither gold miners nor tourists; both groups, he feared, trampled his cherished wilderness disrespectfully. He was a cynic about the growing tourist industry touted by Dall, and often grumbled about the streams of sightseers he had encountered on his numerous trips to Alaska. Tourists aboard a steamer sailing into Glacier Bay in 1890 drew a disgusted mutter from Muir, who wrote "what a show they made with their ribbons and Kodaks! All seemed happy and enthusiastic, though it was curious to see how promptly all of them ceased gazing when the dinner bell rang." Muir was virtually alone in his hostility toward the tourist industry. Henry Gannett called to mind his Yellowstone and Colorado mountain experiences, and thought that Alaska's primary value lay in its potential as a tourist mecca. Gannett later told a news reporter that "when the Alaskan beauties are appreciated and the steamer lines well-advertised, tourists will flock there by the thousands." Perhaps bearing in mind Harriman's own interests in the tourist industry, as well as the scenic boosterism characteristic of F. V. Hayden, who had advertised Yellowstone so well, the scientist extolled the tremendous economic value of Alaskan scenery once its beauty had been as well touted as its gold. "Alaska's grandeur," Gannett wrote optimistically, "is more valuable than the gold or the fish or the timber, for it will never be exhausted. This value, measured by direct returns in money received from tourists, will be enormous."

The botanist Coville suggested that sheepherders should consider the region's unused grassy fields. Alaska's "enormous growth of grass," he noticed, yielded "millions of tons going to waste every year." Fernow, Cornell's forestry expert, also saw "waste" in the unexploited timber all around him, and he wrote that "enormous quantities [of hemlocks] are now going to waste in the forests around Puget Sound, because its [*sic*] value is not known or appreciated in the market." Perhaps it was

the vastness of the Alaskan territory that caused the word
"enormous" to appear so frequently in the vocabularies of the
expedition members. The dominant idea running through their
descriptions of Alaska was one of overwhelming, even bewil-
dering, space that needed to be used, catalogued, and made
efficient. Only Muir seemed to recognize that the development
of Alaska would involve at least a partial destruction of the
scenery that held all the expeditioners in awe.

Leaving Skagway and the newspaper accounts that proclaimed its visit, the *Elder* set off with its passengers to pick
up the scientific party left near Juneau, a town that already
claimed a budding tourist industry. Insulated against the dire
economic struggles around it like some floating medieval for-
tress, the *Elder* plowed its way through the intense gold-rush
dramas. Apparently too absorbed in their starfish, birds,
weather recordings, and bear hunts to note the stark human
tragedies displayed at Skagway, the expedition's scientists, as
they departed, were filled with ideas for the region's further
development. But to the Skagway dreamers they left behind,
the wealth Harriman represented must have only whetted their
appetite for gold.

CHAPTER FIVE

JOHN MUIR'S COUNTRY

At their campsite near Juneau, the scientists who had remained behind to work searched the skies over Lynn Canal on the evening of June 7 for the column of dark smoke that would signal the return of the *Elder*. But the ship did not reach Juneau until after dark to collect the little band of nervous scientists. Then, with all its guests aboard again, the ship proceeded through the intricate network of coastal islands to Glacier Bay, where the *Elder* would anchor for five days.

During the extended stay at Glacier Bay, the scientists hoped to resume their work in earnest. There they could camp on the icy outskirts of the bay, charting and mapping the area with their measuring instruments and collecting specimens of flora and fauna to study back in Washington. The region also lured Harriman with the possibility of killing one of the great bears that, he hoped, roamed freely across the glaciers and their moraines.

When Muir Glacier finally loomed on the horizon the next day, its discoverer proudly surveyed the great river of ice that guarded the head of Glacier Bay. It was familiar territory to John Muir, who had camped for weeks at a time on the massive glacier, exploring its vast icy stretches. Ten years earlier he had built a rough cabin at the foot of the ice, foreseeing long stays alone in the peaceful reverie that the wilderness inspired in him. The expedition steamed straight toward the small cabin and sought anchor close by the great glacier. Dwarfed by the two-hundred-foot-high wall of ice, the ship repeatedly dropped its anchor beneath the green water, but finding anchorage was not a simple process in the deep waters of Glacier Bay. Only after repeated attempts did the anchor hit. As if to roar its disapproval at the intruders, Muir Glacier periodically released enormous ice masses, which crashed down its front in avalanches and broke into the water with a thunderous boom that frightened the Harriman children. Burroughs watched in amazement. It was like a Niagara Falls of ice, he thought. The front of Muir Glacier was a formidable sight, rising up high from the water and stretching for two miles along the coast. Its ice glistened in blues and greens in the sunlight.

Knowing that several of the hunters in his company sought game, John Muir found it a convenient time to tell a favorite story about his hike into a nearby valley some twenty years before. "Howling Valley," Muir dramatically called it, though on maps it was known as Endicott Valley. Once one made the rough trek over eighteen miles of snow and ice to reach the valley, said the naturalist with a gleam in his eyes, all one could hear were the lonely cries of wolves, hundreds of them, lurking in the surrounding hills. But there, at the snowy lake nestled in Howling Valley, one could also find big game, Muir confided impishly. He knew that the ears of the frustrated hunters would perk up at the hint of bears in the valley. He was not mistaken. Harriman immediately sent six packers off, carting sleeping bags and camping equipment in the direction of Howling Valley. In an hour the tycoon planned to follow them, with Merriam, Grinnell, the two physicians, Trudeau and Morris, and Captain Luther "Yellowstone" Kelly, a former scout for Custer, whom Harriman had invited on the ex-

Reid Inlet—Glacier Bay, by Grove Karl Gilbert.

The Reid—Glacier Bay, by Grove Karl Gilbert.

pedition. Armed with Winchester rifles, the six men struck out in late afternoon across the glacier for an overnight hike into the wilderness.

Muir watched them trudge off into the distance, remembering his own solitary hike to the valley in 1879, accompanied only by the faithful little dog Stikeen. He had remained silent, for some reason, about how difficult it had been then to find his way over the glacier. In fact, his own journey out to Howling Valley had nearly ended in disaster, but perhaps the Scotsman was too proud to tell his company that he had gotten lost in the confusing maze of icy crevices that marked the route to the hunters' destination. Twenty years earlier, when Muir had camped with his old friend Hall Young at the foot of the glacier, he had struck off on his own across the ice. By nightfall, a worried Young had built a large fire to signal his friend, and finally sent out Indian guides with torches to search for him. The Indians found a weary Muir, who staggered into camp around midnight with them and sank to the ground. "The old glacier almost got me this time," he had sighed. It was a confession that Muir would not readily make. The glacier had been riddled with deep, hidden crevices in which a man could be lost forever—entombed in the ice. In 1890, Muir again brushed with near disaster in Howling Valley when wolves prowled nearby. His alpenstock was his only protection against the wild animals, but, fortunately, the wolves never attacked. The Harriman hunting party, blissfully unaware of the possible dangers that awaited them, set off across the icy terrain with their guns.

The launch that carried Harriman's party to shore returned for several more expeditioners who were ready to explore the rocky moraine left by the glacier when it receded. Curtis, not wanting to wait for the launch, set out on his own in a canvas canoe loaded with his photographic equipment. All along the shore, the explorers found boardwalks stretched out over the rougher areas of the moraine, and the men surmised that a steamboat company had probably placed the planks there for the benefit of the tourists it brought annually into the bay.

Already, the remote reaches of the wilderness Muir had discovered only twenty years before bore the marks of a tourist industry eager to bring in sightseers. The land the glacier exposed as it retreated was too hazardous for the average tourist. In fact, Burroughs and Gannett found it difficult to cross one of the swift, icy streams, and Dellenbaugh, clad in high rubber boots, made two trips through the current to carry the men across on his back. Muir's cabin lay farther ahead, and they approached it with curiosity, finding a simple little structure that had weathered the years well. Evidence of other campers lay strewn about the room. Apparently, Muir's wilderness abode had served as a way station for other adventure seekers.

Wandering out of the tiny cabin, Gannett, Gifford, Burroughs, and Dellenbaugh looked up at the glacier just as a tremendous ice mass broke from its front and hit the water with an explosion. A jet of water shot halfway up the glacier and an enormous wave rolled out into the bay. It was a fear-

The Way of the Transgressor—Glacier Bay, by C. Hart Merriam.

some sight. As Burroughs put it, "We saw the world-shaping forces at work."

All that night the noise of falling ice shattered the sleep of the passengers on the *Elder*, and the great ship rocked periodically in the swell of rough waves. The restless night, however, did not prevent the expeditioners from rising early to scout the new territory. Gannett set out in a launch to reach the west shore and chart the front of the glacier, while Dall set signals on the coast for him. But the ice proved too dense for Gannett's small launch, and the two men gave up their ideas of work. Instead they joined Mrs. Harriman for a hike on the glacier. Dellenbaugh also found work impossible. The cold wind on the shore made his fingers too numb to hold a paintbrush and he wandered off alone to make photographs for later sketches. Muir, meanwhile, accompanied the Harriman daughters, Mary and Cornelia, their cousin Elizabeth Averell, and Dorothea Draper, all of whom he had labeled "the Big Four," on a three-mile hike along the glacier. Blue ice grottoes suggested a magical land to the girls, and Muir was pleased that "his" glacier had cast its spell of enchantment on his young company. But at the front of the glacier, the "fairy-tale" ice had turned dangerous: An enormous iceberg rose suddenly to the surface of the water, breaking off from the submerged reaches of the glacier. Waves crashed against the shore, and the Harriman family party scrambled to higher land to avoid the crest of a wave headed directly toward them. Far below them, Curtis and his assistant, D. J. Inverarity, faced the brunt of the watery assault in their flimsy canoe.

As the bear-hunting party straggled back over the moraine after their long trek to Howling Valley, they could hear the roar of the water. From afar, the men watched the dramatic birth of the iceberg in fascination. But their awe turned to horror when the men spotted the tiny canoe bearing Curtis and Inverarity near the glacier's front. The two men paddled furiously as wave after wave tossed them about violently. Loaded with heavy camera equipment, the boat thrashed about

Glacier, by E. H. Harriman.

in the water, and Merriam looked away as the largest wave descended on the boat, thinking the photographers were surely doomed. But Curtis and Inverarity headed the canoe directly into the wave and, to the amazement of the onlookers, they managed to ride up over its crest. Still, danger surrounded them as sharp chunks of ice rained down, and they paddled rapidly to escape the glacial fragments. Years of living on the Seattle coast had trained the two men as expert canoeists, and they astonished the rest of the expedition when their small craft finally scraped the shoreline and they emerged from its confines, shaken but unhurt. The glass photographic plates that Curtis had tucked into the canoe had been ruined, however, and he sadly began picking them out of the battered boat.

The photographers' near-disaster was not the only dangerous experience in the first two days at Glacier Bay. The Harriman hunting party wearily marched back to the *Elder*, exhausted by their night on the ice and a futile search for game. Grinnell's new hiking boots had rubbed blisters on his feet and the unaccustomed exertion had taken its toll on the entire party. But Merriam had fared the worst. He limped along the rocky beach, his knees stiff with arthritis and his feet bruised by the sharp ice. When told by the packers of Merriam's condition, Cole and Kearney quickly hiked out to meet him and take his twenty-pound pack from him. Muir, who perhaps felt pangs of guilt at the sight of the old scientist hobbling in the distance, hurried to help him back to the *Elder*. Once on board, the tired hunters ate a quick supper and fell into their beds.

It had been a harrowing twenty-four hours for the hunters. After departing from the *Elder*, they had crossed the rocky moraine for several miles, slipping on masses of partially buried ice before climbing on the glacier itself, where they proceeded until the darkness hindered their passage. The men stretched out their sleeping bags, after cooking the bacon brought by the packers, and tried to rest on the hard bed of ice, but the cold crept up from the glacier and penetrated their sleeping bags. After three hours of discomfort, the men gave up on rest and trudged ahead across the ice. Freezing rain

Before the Great Berg Fell, by E. S. Curtis.

poured down on their heads when they finally reached the snowy edge of Muir Glacier which, according to Muir's story, indicated that Howling Valley lay not far over the next ridge. Sinking to their knees in melting snow, the men fought their way through the last portion of the journey, keeping a watch for animal tracks in the snow. But one of the packers had had enough. The deep snow proved too much for the experienced guide, and he turned his back on the foolish venture. The indefatigable Harriman refused to give up, however, until the party had at least seen the mysterious valley. When they began to come upon the deep, hidden crevices that had tormented Muir years earlier, the party roped themselves together, fearing that a man could sink irretrievably into a bottomless glacial crevasse. They found a disappointing sight when they reached the divide that looked down on Howling Valley. What must have been the lake lay covered with icy snow, and the valley stretched out in silent whiteness, its quiet unbroken by any trace of game. The animals, suggested a tired Merriam, must still be along the coast. Slowly, the men retraced their steps and headed back to the comforts of the *Elder*. Harriman's bear would have to be found elsewhere. Hearing the story of their failure, John Burroughs wondered slyly whether it was John Muir's imagination that had done "all the howling."

Albert Fisher was tired of the endless gazing at the glacial ice and he looked forward to the next day, when the *Elder* would drop off parties to hunt their specimens and explore the land. Point Gustavus, Fisher thought, would be an ideal place to set up camp. The following day, while Curtis and Gifford remained at Muir Glacier to work, the ship puffed along the bay to set loose its scientists on the land. Early in the morning, Ritter, Coe, Coville, Saunders, Kincaid, and Trudeau requested that they be let off ship to do some dredging for marine life. Merriam watched the shoreline ease past his stateroom window as he lay confined to his bed for the day, his knees and feet still swollen and sore from the Howling Valley fiasco. After breakfast, he watched as Fisher, Fuertes, Ridgway, Kearney, Fernow, and Cole took off enthusiastically at Point

"The Two Johnnies": Burroughs and Muir.

B. K. Emerson and Grove Karl Gilbert.

Tent and Child at Glacier Bay.

Sealers' Camp in Glacier Bay, by E. S. Curtis.

Gustavus, carrying camping equipment for three days. In the lush, mossy woods, Fisher hoped, they could acquire a large collection of birds.

The Pacific Glacier appealed to Muir's instincts, however, and along with Charles Palache and Gilbert, he asked to be landed far up the bay, where he would lower a rowboat into the water and explore the glacial shoreline for three days. Dellenbaugh briefly entertained the thought of joining the small party, but Muir's adventurous, indomitable spirit for exploration finally changed his mind. It was too hard to tell, the artist thought, what sort of "chase" Muir would lead him on, and he watched as the three men paddled their way among the icebergs toward the Hugh Miller Glacier. Ice packed the water too densely for the *Elder* to pick its way to the glacier, and its iron propeller, Captain Doran knew, could easily snap if it met a large iceberg beneath the water.

Late in the afternoon, south of Berg Inlet, the *Elder* picked up the dredging party headed by Ritter. As the ship approached shore, curious Indians paddled up beside the *Elder* to meet the visitors to Glacier Bay and display the trinkets that the white travelers often seemed interested in buying for souvenirs. Dellenbaugh bartered with the natives for one of their paddles and a rattle. Ritter's party had already met the group of natives during their brief stay on the shore. At lunchtime, the scientists had encountered the Indians' camp, where they had eaten with the natives, who graciously shared gulls'

Sealer's Camp (note Merriam cropping line).

eggs, boiled marmot, and hair seal with their guests. The explorers confessed they looked forward to a more "civilized" dinner on board the *Elder*.

Since Doran was confident of making anchorage at Muir Glacier, the ship steamed back to its previous site for the night. Curtis and Gifford greeted the expedition happily, glad for a chance to warm themselves on board. Muir's cabin at the glacier had not provided much shelter from the cold. Merriam noticed an odd change in the front of Muir Glacier and the two men explained that during the day the largest iceberg yet had fallen off its front, creating a wave that splashed nearly to the top of the glacier. The waters beneath Muir Glacier, Curtis now well understood, could be quite hazardous, and he was glad that he had watched the birth of the iceberg from the shore this time.

Back at Point Gustavus, the Fisher party had also befriended the natives they met. In the evening, the Indians approached the scientists' camp, tucked in a little glade a brief walk from the shore. Though Fisher refused the salt the Indians offered him for sale, they remained at the camp to look at the numerous bird specimens strewn about the tents. Soon the natives were teaching Fuertes and Fernow the Indian names for the various birds. Later in the evening, the campers sat huddled about a large crackling fire while Fernow, a German who had fought in the Franco-Prussian war, entertained his company with stories of the battles. A sleepy Fuertes felt cozy in the warm firelight and content with his new friends; it was good to be here, he thought, as his company settled in to sleep before the next day's hard work.

The eleventh of June dawned bright and sunny. At Hugh Miller Glacier, Muir, Palache, and Gilbert woke to find magnificent reflections of the Fairweather Range in the still bay. It would be a lovely day, Muir thought, to do some exploring along Reid Inlet. Back on the *Elder*, Merriam was coaxed out of bed by the brilliant sunshine and, though he was still stiff, a cruise among the scattered islands in the bay appealed to him as he joined Harriman in the steam launch. It would be

a day of leisurely sightseeing in the launches for most of the
Harriman family, and they planned a picnic on one of the
islands. But the Harriman daughters preferred to spend the
day on the glacier and, along with Gannett, Dellenbaugh, and
Brewer, they steamed toward the shore. Inverarity met them
with planks to help the girls walk ashore in ladylike fashion.

While the other scientists scattered in various directions to
work, Burroughs and Keeler, both feeling at odds with their
scientific companions, climbed the mountain that rose behind
Muir's old cabin and took in the grand view that spread beneath them from the summit. From their vantage point three
thousand feet above the glacier, the two poets drank in the
panorama beneath them. To the north, Muir Glacier extended
into the distance like a vast prairie of ice, and to the west the
Fairweather Range rose against the horizon. The snowy peak
of Mount Fairweather stood regally under a blue, cloudless
sky. To the florid Burroughs, the place suggested a "solitude
as of interstellar space." Walking a short way from his companion, Keeler pulled out his notebook and penned a hasty

Ground New-Made by Hugh Miller Glacier, by Grove Karl Gilbert.

Forest Felled by LaPerouse Glacier, by Grove Karl Gilbert.

Front of the Hubbard Glacier, by Grove Karl Gilbert.

description of the place. There was, he thought, "a latent and terrible power underfoot." All he could hear was the "drip of melting ice as it falls into the crevasses." The tracks of a mountain goat were the only signs of life on top of the mountain. But in the bay beneath the two men, several steam launches crisscrossed the water and the *Elder* floated like "a little toy boat . . . amid cakes of ice," a reminder of the human presence in the grand, desolate surroundings.

In fact, the scene below fairly crawled with scientists. At the tide line, Ritter and Saunders studied the visible marine life while Kincaid and the taxidermist Starks explored the glacier. On his hands and knees, Kincaid pried apart tiny crevices with an ice pick, seeking insects and the strange brown "glacier worms," which, he knew, "had been known for many years to the Alaskan pioneers but had apparently never been called to the attention of scientists." Across the bay, Coville roamed over a site where the glacial movement had buried entire trees. Gannett, with the help of the Harriman girls, set up his plane table on a glacial "island" to measure the icy mass, while nearby Dellenbaugh sketched a distant mountain. It was, he assumed, the famous Mount Fairweather everyone was talking about.

For the campers on Point Gustavus, it had been an idyllic Sunday, which the ornithologists had spent chasing birds and watching whales cavorting in the bay. Fuertes had hiked far back into the woods that afternoon, tempted onward by a strange birdcall. "It seemed," he reflected later, "like some Grimm's fairy-tale bird, never seen, but heard, and luring the child on and on." Indeed, he thought the woods seemed so enchanted that one virtually expected a witch's hut to appear. Camping in the wilderness was awesome to the young artist, and, in camp that night, he wrote home that the surrounding beauty was inexpressible. Fuertes already felt sorry about rejoining the *Elder* in two days. Life aboard the luxurious vessel was too soft, he resolved, and he longed to spend a week or so "roughing it" in the woods.

The aggressive, businesslike Fisher was displeased with the number of specimens his party had collected at camp. He worried that they would not be able to spend enough time ashore to gather more during the voyage. He had come to the region to work, not to socialize, and he held high expectations for the expedition's findings in ornithology. The forty-five mammals and twenty-five birds (which included several shrews he thought to be rare specimens) accumulated at Point Gustavus did not even approach his goals.

If the party had been interested in collecting mosquitoes, however, there would have been no shortage of specimens. Mosquito netting had somehow been left behind on the *Elder*, and the scientists woke in the mornings to find their faces covered with bites, a condition that did not lessen Fisher's grumpiness. Fuertes, despite his affinity for "roughing it," discovered one morning that he could barely open his eyes for the bites that covered them. He spent a miserable hour holding up one eyelid with his thumb before the swelling went down.

Muir's party on Hugh Miller Glacier had enjoyed a more rewarding camping excursion. During their days among the glaciers, Palache, Gilbert, and Muir had discovered that the Grand Pacific Glacier had split into three smaller ones. In the excitement of their finding, they named the largest in honor of Edward Harriman. Now "Harriman Glacier," measuring nearly a mile wide, stood where the old Pacific Glacier had melted. The Muir party, like the ornithologists, found friendly Indians approaching their camp. Two natives generously shared their meal of gull eggs and wild celery with the scientists before they hiked on, carrying their bark canoe over their heads.

But, like Kearney back at the ornithologists' camp, Gilbert and Palache were anxious to rejoin the expedition. After three clear days on the ice, they were badly sunburned and the two men looked forward to meeting the ship at the Indian village they had designated beforehand. On the appointed morning, Muir, Gilbert, and Palache rose early to make the hike, but, when icebergs hindered their way, they soon realized it would be impossible to meet the *Elder* on time. Spotting the steamer as it puffed by below them several hours before the scheduled meeting time, Palache and Gilbert waved in vain to hail the ship from their position on top of an iceberg. A bemused Muir coolly watched his friends' frantic efforts. Possibly unaware of Gilbert's wilderness experience, which was at least as extensive as his own, Muir wrote that his colleagues were "wild to get on the steamer." His own long days enjoying the solitude of the wilderness made him impervious to any fear of being left behind. The tight schedule maintained by the *Elder* rubbed against his independent grain. After the expedition had ended, Muir wrote his old friend Hall Young that he "longed to break away from the steamboat and its splendid company, get a dugout canoe and a crew of Indians and, with you as my companion, poke into the nooks and crannies of the mountains and glaciers which we could not reach from the steamer." Three days in camp on the glaciers had not been enough for the impassioned explorer.

To the relief of Palache and Gilbert, the *Elder* sent two launches to search for the "lost" party when the three men failed to show up at the Indian village. Harriman himself steered one of the boats, and he threaded his way through the ice to reach the stranded party. According to Merriam, the railroad baron thrived on leading such risky endeavors. "Running the launch," Merriam wrote, "was one of Harriman's favorite diversions. The greater the danger, the surer he was to take this post." Harriman heroically rescued his guests, and they were soon aboard the ship nursing their sunburns. Muir enjoyed an attentive audience when he lectured on his party's discoveries in the evening. His listeners greeted the announcement of the new Harriman Glacier with loud cheers, sending up the "war cry" of the expedition: "Who are we? Who are we? We are, we are, the H.A.E.!"

CHAPTER SIX

"CRADLED IN CUSTOM"*

In a heavy morning rain, the *Elder* approached Sitka after steaming south down Peril Strait all night. After mooring offshore, the expedition members motored over to the town in launches, shielding themselves against the downpour with their rain slickers. The outline of a small town soon came into view, its rooftops framed by the spruce-covered mountains rising gently behind them. Low barracks left from the days of Russian settlement ran along the streets, and the spire of a wooden church rose elegantly above the town. Once on shore, the group sought shelter from the rain in the Alaska Commercial Company's store, whose counters were lined with Indian trinkets for the tourists who descended on the town whenever a steamship docked. Ironically, the Indians preserved their cultural artifacts by making them for tourists, whose presence contributed to the natives' cultural decline. Despite the high prices, several scientists bought souvenirs, which the company bought from the Indians and sold to customers at a healthy profit.

Here at Sitka, some thought, was a lively pocket of "civili-

*From "The Call of the Wild," in *The Spell of the Yukon* by Robert Service.

Sitka, by E. S. Curtis.

zation" in the midst of a wilderness otherwise populated by coarse towns. The capital of Alaska, Sitka was inhabited by an educated, society-conscious elite who had moved to the site when the United States had bought Alaska in 1867. Once the locale of licentious, free-booting Siberian fur traders who carried off the Aleut women and enslaved or killed the men, Sitka had eventually become the Russian American Company's North American headquarters and had settled into a sedateness symbolized by the onion-domed church. American settlers tended to follow suit; they were a far cry from the Skagway "howlers." The Slavic buildings still marked the town's earlier era but newer, American-built structures cropped up throughout the town in an architectural style that Charles Keeler thought utterly "hideous." It was shameful, he thought, that the ugly American buildings destroyed the historic charm of Sitka. Once the Americans came to dominate the town, the Indian residents, who had survived the early "promyshlenniki" atrocities and who now made up about half of the town's population, had been pushed to the outskirts of Sitka on what was called a "ranche." Dellenbaugh, Keeler, and Ritter toured the small houses of the Indians, ignoring the furs and baskets thrust at them by hopeful native peddlers. The strange little

Old Russian Church—Sitka, by E. S. Curtis.

cabins, they discovered, were only a sort of camp inside, with beds and traps scattered about a stove situated in the middle of the cabin. The animal grease used by the natives produced an overpowering odor in the houses, and when the three men entered one, they recoiled simultaneously at the noxious smell. Pictures of Christ, they noticed with satisfaction, hung on virtually every wall of these Indians' houses.

Merriam visited with his old acquaintance George Emmons, a retired naval lieutenant who had sailed with Wilkes on the great United States Exploring Expedition of 1838–1842, which had first confirmed the existence of an Antarctic Continent. Emmons, one of the last survivors of the Old Navy, had subsequently pursued a long career in scientific study of Alaska. Now he had retired to remote Sitka. Modestly neglecting to mention that he had spent most of his life in polar wastelands, he explained to an incredulous John Burroughs that Sitka had the most pleasant climate he had ever found. Under assignment by the Navy, Emmons had studied the Tlingit Indians of the area for years, collecting their artifacts and selling them to East Coast museums. He had gained the natives' trust through the years and established a long-standing friendship with Chief Tentlatch of the Tlingits. Emmons' rapport with the Indians fascinated George Bird Grinnell, whose own studies, of course, had brought him into close contact with many western tribes, and he struck up a quick friendship with Emmons. Asked about the area's big game, Emmons took the men to visit an Indian who had recently killed a huge brown bear and carefully saved the skin. Merriam bought the bear's skull from the hunter to add to his collection of mammals.

In the evening, the Russian church opened its doors to the town's new tourists, and the Harriman guests poured inside to inspect the lavish interior. Huge columns stood in line across the center of the church, with pine boughs swung between them. A high altar ran across one entire wall, covered with oil paintings—icons that looked strange to the eyes of the visitors. Beaten silver had been worked into the paintings of religious figures until only the faces were visible in paint. Off in a little alcove to one side, a painting of the Virgin Mary hung, all but her face shining with the pounded metal.

Afterward, Governor Brady, Lieutenant Emmons, and other notables from Sitka boarded the *Elder* to dine with the Harriman party. As servants uncorked champagne bottles and wound up the graphophone for the gala affair, the conversation soon turned to the fate of the Sitkan Indians. Separate churches and schools had been set aside for natives since the days of Russian settlement, and the practice had been continued under the Americans. The Indians, deemed to be savages, then stood lowest in the racial hierarchy set up by the whites who had overrun the natives' ancient culture. Even so, William Dall insisted, the Indians fared far better under the guidance of American missionaries than they had under Russian governance. George Emmons lightened the discussion with tales of the local Indians' immense courage and their great hunting skills.

Brady and his wife joined the Harriman party again in the morning to cruise down to a hot springs located some fifteen miles south of Sitka. The town's wealthier citizens often frequented the hot springs; the men's club, the Sons of the Northwest, found it a convenient spot to relax on their hunting and fishing outings. Along the way, a large island looked promising to the hunters on board, and Grinnell, Merriam, Gifford, and eighteen-year-old Mary Harriman took off on their own, armed with their rifles and two scrawny dogs from Sitka, determined to track down the legendary big game of the area. At the hot springs a few miles farther on, steam puffed ominously from the earth and warm water bubbled up, smelling strongly of sulphur. John Burroughs cautiously took a sip of the strange water and imagined himself in a scene out of Hades. It would be handy, he thought, to have such a hot spring on board the *Elder* to ease the Arctic chill that seemed permanently embedded in his bones. But John Muir felt sickened at a scene near the springs. The keeper of the hot springs had "murdered a mother deer and threw her over the ridgepole of his shanty, then caught her pitiful baby fawn and tied it beneath its dead mother." Muir fled the pathetic scene and spent the hours wandering about the island's woods gathering flowers.

But the naturalist could not escape another deer "murder," this time at the hands of his colleagues. When the ship stopped to pick up the hunting party, a slain doe lay sprawled out before the group. The small party credited Mary Harriman with shooting the deer, but, embarrassed by all the fuss, she modestly insisted that Merriam had killed the animal.

It had been a fruitful trip for the marine collectors as well as the hunters; Ritter and Saunders found starfish in abundance on the beaches and Coe had gathered up a long red worm that stretched to nearly ten feet in length. The shorelines, it seemed, fairly crawled with the strange, ropelike worms. Fuertes had discovered the area to be filled with birds he had never before seen alive, and he covered page after page in his drawing pad with colorful sketches that won praises from his older colleagues.

One full day remained at Sitka before the expedition turned to the open ocean to head for Yakutat Bay, and George Bird Grinnell put the time to good use by visiting with his new friend George Emmons. A rare treat awaited the ethnologist when he and Charles Palache knocked on Emmons' door that morning. Recognizing his two guests' compassion for the Indians, Emmons escorted Grinnell and Palache to the house of Chief Tentlatch to see some secret carved headdresses made from the killer whale for sacred tribal rituals. Only the trusting friendship between Emmons and the chief allowed the two strangers to view the elaborate secret possessions. Merriam, off exploring the Indian "ranche" on his own, missed the private exhibit, but he was delighted to find an extremely rare medicine man's pot, which he quickly bought from its owner. Fernow and Devereux bargained with the Indians for their animal furs, and both came away from the village weighed down by the large skins of brown bears. If they could not procure the fur by their own hunting, at least they would have a souvenir of the great beast they sought.

Dellenbaugh, though, had already satisfied his curiosity about the Indians on the day he arrived and did not care to visit the smelly village again. Instead, the artist headed off alone into the spruce woods to sketch and explore. Following the mossy banks of the Indian River into the woods, Dellenbaugh came across John Muir, who was busily pulling down spruce boughs that bore a rare "hermaphrodite" blossom. The flowering of the spruce trees was at its peak in mid-June, and Muir examined the green and red blossoms in detail, pointing out to Dellenbaugh the vast differences in their sizes and colors.

After a leisurely day, Muir and Dellenbaugh found their way back to town to attend a reception given in the Harriman expedition's honor by Governor Brady. Muir turned his nose up at the stilted gathering: such a "society affair . . . looked queer in the wilderness." But similar gatherings were far from unusual in Sitka. The town's white population enjoyed a social life enhanced by the Sitka orchestra, a theatrical association, and numerous lively balls that lasted into the early hours of the morning. As the last stop of the steamship tours up the Alaska coast, Sitka had acquired a small reputation for its cultural activities among the tourists, who celebrated the end of their northward voyage at the town.

Harriman, of course, was no ordinary tourist. For such a prestigious visitor as the tycoon, Governor Brady presented a dignified, formal affair. A houseboy led the guests upstairs to the main room of the governor's official house, where the party dutifully passed through the governor's receiving line. Dellenbaugh, feeling as if he were "running a gauntlet," exchanged superficial pleasantries with his host. Charles Keeler felt "peculiar" dressed in his white shirt and coat for the evening; usually his old comfortable clothes were all he needed in the wilderness. He had written enthusiastically that "there is no snobbery or conventionality about [the expedition members]. They are all gentlemen and behave as such so we all feel very much at home however we may be dressed." But amid the "conventions" and fancy clothes at Brady's reception, Keeler shifted about uncomfortably. Harriman had planned rousing entertainment for the evening, which helped relax Brady's formal cake-and-coffee reception. The railroad baron had invited several of Sitka's Indians, including Chief Tentlatch, all sporting white shirts and jackets. In the governor's mansion, Harriman set up his graphophone to record the Indians' songs and their language. The natives spoke and sang into the strange machine with no trace of shyness. In a few minutes they were stunned to hear their voices spinning back from the wax record as gustily as they had sounded originally. Governor Brady himself, fascinated by the process, volunteered to make a speech into the strange gadget. When the excitement over the graphophone died down, William Brewer climaxed the festivities with a talk on climate, a suitable topic, he slyly averred, for the residents of "sunny" Sitka.

But a more exciting event upstaged Brady's reception that night. The arrival of the tourist ship *City of Topeka* promised mail for the expeditioners, and they left the reception hoping for news from their families. When the *Topeka* docked later in the evening, swarms of tourists poured across the Sitka streets and into the curiosity shops, giving them a thriving business. It looked like a revival, thought a cynical Muir, as he and Keeler met friends from Berkeley and San Francisco who were vacationing on the ship. Keeler's friends, however, bore bad news for the writer. A manuscript on birds he had recently completed had been poorly received by his publishers. Worried and angry over the latest turn of events, he considered leaving the expedition altogether and boarding the *Topeka* to return home and resolve his business problems. After much thought, Keeler finally decided to remain with the *Elder*. Balancing a notepad on his camera, he stood on the Sitka pier and hastily jotted a quick letter to his wife, giving her instructions about handling his publishers. Keeler was not alone in his desire to return home; John Burroughs wrote his son, "I should be quite content to go home now." Burroughs worried constantly about his grapes back in New York and news that the East Coast had suffered a heat wave disturbed him. Dreams of his grapes withering under the sun filled his sleep, and his letters home anxiously probed for word of his crop. But Sitka had also pleased Burroughs—here he had discovered among the educated populace four people who had read his books. Always a self-promoter, Burroughs introduced himself in each town he explored, keeping count of the people who recognized his name. The number of fans he found in Sitka gave him a favorable feeling toward the town.

Despite Burroughs' and Keeler's temptation to return to the States via the *Topeka*, the Harriman expedition remained intact, and the next day they bade their farewells to Sitka. Harriman was sitting on the porch of Brady's mansion, giving the Indians a last chance to hear themselves on his graphophone, when he was startled by the sight of an Indian brass band marching toward him. The band struck up the same tunes that the Indians had sung into the recorder the night before, and when they finished, Harriman responded by replaying the songs to them. The *Elder*'s sharp whistle called its passengers to board and, as the daily rain began to fall, the expeditioners hurried to embark. The Indian band reassembled on the pier to salute their departing visitors, and though rain drenched them, the native musicians enthusiastically played lively renditions of patriotic American tunes: "Yankee Doodle" and "Three Cheers for the Red, White, and Blue" followed the *Elder* across the dark waters as it steamed farther and farther away from the little band of "savage" redmen heroically puffing their musical farewell.

CHAPTER SEVEN

MALASPINA'S MISTAKE: YAKUTAT BAY

Albert Fisher waved good-bye to the Sitka group on shore, feeling relieved that the ship was headed to new Arctic sites; now they would cruise northward in the open ocean, where tourist boats never ventured. In five days at Sitka, Fisher's bird collection had barely increased, and, since the expedition-ers were never told more than one day in advance what the *Elder*'s schedule was, he had no idea what future opportunities might hold. The next stop, rumor had it, was Yakutat Village and Bay, and he hoped the new anchorage would yield more profitable finds. It was a long day's travel to reach Yakutat, which lay farther up the Alaska coast, and, for the first time, the ship would leave the coastal channels for the Pacific Ocean. John Burroughs dreaded the thought of the ocean's rough waves and the inevitable pitching of the *Elder*. Seasickness plagued him, even in the milder coastal waterways, and he had already spent a day or two in bed suffering the dizziness and nausea brought on by the rolling of the ship. But the ocean proved so gentle, wrote the naturalist, "that it did not disturb the most sensitive," meaning, of course, himself.

Impenetrable, forbidding-looking forests lined the coast in the distance, and an occasional glacier cut a jagged edge through the dense growth. The rolling green mountains and pleasant spruce forests of Sitka had given way to what Dellenbaugh called "the most forbidding country" he'd ever seen. But the ice and rocks did not deter Harriman when he spotted a glacier that caught his interest. La Perouse and Brady glaciers appeared to Gilbert to descend straight to the sea. Harriman ordered the *Elder* closer for a better look at these magnificent sights. Taking several scientists along, he set out in a large boat, towing a smaller one in case of emergency, to explore the face of the glacier. It was a bedraggled party that soon returned. One boat had sprung a leak on the sharp icy rocks at the shore and freezing water drenched everyone. Harriman clutched one of his hands tightly in the other, trying to stop the bleeding from cuts he had received from a sharp nail on the boat.

Glaciers continued to rise steeply along the shoreline, but the party had had their fill of exploring the harsh areas on foot. For the remainder of the day they contented themselves with watching the glaciers from the comfortable deck chairs on the *Elder*.

Most of the ship's passengers slept as the great steamer finally docked at Yakutat Village around midnight. The small village stood at the entrance of the large Yakutat Bay, which was lined with enormous glaciers dominated by the immense Malaspina, which ran along the ocean for fifty miles. In the past two years, Malaspina Glacier had been the site of several gold-mining disasters when naïve prospectors attempted to cross its vast stretches of snow-filled canyons and mountains to reach the Yukon's promised land. Those who managed to survive the months of travel often reached their destination deranged by the torturous journey, or snow-blinded by the intense glare of the sun. Even as the Harriman party pulled into the bay, groups of the desperate miners were making their way slowly along the glacier, many miles from their point of departure at Yakutat Bay. But the members of the *Elder* were oblivious to the harsh battles for survival being waged on the glacier that they admired in the morning light.

Yakutat Village, they thought, was "charming"—a mission settlement of more than one hundred residents with a Swedish missionary to "look after" the Indians. Three of the expedition's scientific parties had set their sights on the opposite side of the bay, near the imposing Malaspina Glacier, and soon the *Elder* steamed across to deposit a group of scientists and hunters at the moraine along the shore—rumors at Yakutat Village indicated that bears often prowled the tree-lined shore at the glacier's edge. Morris, Devereux, and Grinnell marched off with their rifles in one direction, while Averell, Trudeau, and Kelly set off along the beach, finding bear tracks to follow. A party headed by Merriam pursued lesser game. The scientists set out their traps and, by nightfall, Merriam, Palache, and Fernow had captured twenty-two mice. For three days the intrepid parties explored the edge of the Malaspina Glacier, tucking their camps into the sand flats and willows that stood between the glacier and the bay.

While the parties at the Malaspina busied themselves on the glacier's edge, the *Elder* steamed farther into Yakutat Bay into a narrow inlet called Disenchantment Bay. As the boat waited to push forward through the thick ice floe that blocked the bay, Tlingit Indians paddled alongside the *Elder*, holding up their furs and skins for sale. Harriman invited the new visitors on board and struck up a friendship with an Indian named James, who sported a colorful red patch over his left eye and wore a jaunty, broad-brimmed felt hat. Soon Harriman was entertaining his native guests with the graphophone recordings he had made of the Tlingits at Sitka, which fascinated the Yakutat Indians. With his piratelike appearance and his detailed information on the area, James impressed Harriman so much that he invited the Indian to join the expedition as a consultant for the pilot. "Indian Jim" soon became a familiar, well-respected figure on board the ship. Meanwhile, the ice had opened enough for the *Elder* to wedge its iron hull into Disenchantment Bay, where the vast Hubbard and Dalton glaciers lay. Through the clouds, Mount St. Elias rose high above the narrows, dwarfing the other mountains in its range. To the right of the *Elder*, a cluster of about forty Indian tents stood in a small village, each structure with a boat tied before it. It was odd, thought Dellenbaugh naïvely, to see Indians without horses, but in the rough icy region of Disenchantment Bay, a boat was the only useful means of transportation. Though the bay extended only two miles in width, its waters stretched deep beneath the ship. At dinner, Captain Doran boasted that the *Elder* was the first ship ever to enter Disenchantment Bay—only parties in small launches had attempted to explore the narrow inlet to its end. Dall then related the origin of the bay's name. When the Spanish explorer Alejandro Malaspina searched for the Northwest Passage in 1791, Dall explained, ice had blocked his passage at the entrance of the bay. His hopes for the ultimate sea passage dashed once again, the Spaniard had named the icy bay stretching out in tantalizing fashion before him "Disenchantment Bay." Years earlier, Dall himself had named the Malaspina Glacier in honor of the courageous explorer who had navigated the rough Arctic waters in his vain search for the elusive Northwest Passage. Although the *Elder* made history as the first steamer to venture forward into the narrow passage, anchorage proved almost impossible in the deep waters. After several unsuccessful soundings to determine a proper spot for dropping anchor, the crew finally turned to the one-eyed Indian Jim, who calmly pointed out the next place to try. Harriman had made a wise decision in bringing along the native; this time the anchor held.

Disenchantment Bay held its fair share of enchanting sights for the Harriman Expedition. The group decided en masse to leave the "dis" off the name of the bay, agreeing that its tree-covered mountains and flowering islands were worthy of a more appropriate name than that provided by Malaspina a century ago. As the *Elder* neared Egg Island in the middle of the bay, the smell of its wildflowers grew so intense that when Muir breathed deeply he found he could inhale the scent drifting across the water from a distance of half a mile. For John

Burroughs, though, the scene seemed "weird," with birds and flowers incongruously in abundance among the desolate bay of "savage" ice. It seemed, thought the naturalist, like "a special playground of the early ice gods." He fantasized strange faces and animal shapes in the rocky forms that loomed along the narrow sides of the fjord. The safe haven of the comfortable *Elder* appealed to Burroughs far more than the gravelly moraine along the shore, but most of the Harriman expeditioners separated into small groups and set up campsites. For five days in the area, the scientists and vacationers climbed, explored, and cruised in their launches beneath the cloudy skies of an Alaskan June.

Gannett and Dall ran base lines along the glacier and produced plane table surveys of the region, mapping it for the future, while they taught Cornelia Harriman and Elizabeth Averell the rudiments of their professions. The two girls helped the scientists with their measuring instruments on the glacial moraine, daintily lifting their long, cumbersome skirts to hike across the rough ground. Indians, intrigued by the strange instruments, studied Gannett from a respectful distance. His scientific rituals looked like some weird ceremony and always provoked interest from the Alaskan natives. Inverarity, blessed with a boundless youthful energy, carried his camera equipment to shore at Disenchantment Bay while Curtis remained behind at Malaspina Glacier so that the photographers could cover the entire spectrum of the bay. Gilbert and Muir, who always sought the most remote areas to explore, set off to investigate the far side of the bay. When rumors of bear tracks filtered back to the *Elder* from one of the hunting parties, Harriman picked up his rifle and took a launch to seek the elusive creatures. Kelly had been so near a bear the day before that he could actually smell it. For an entire day, the railroad tycoon marched through the moraine for signs of the animal. Late at night, he returned to ship, tired and discouraged, toting a rifle that had not yet even been fired.

Hunting seal would have been a far easier sport for Harriman, had the seal furs held the same prestige as a bearskin.

Indian seal hunters in their dugout canoes quietly glided over the waters of Yakutat Bay, seeking the prey that provided them with warm clothing, food, and fuel. The Harriman Expedition had arrived at the time of the Indians' annual seal hunt, and it was a busy event on shore as well as in the water. At the Indians' campsite on the bay, the Harriman parties assayed the Indians' bark houses and crude tents while Indian women and children skinned the seals caught by the men. It was, thought a horrified Keeler, "one of the filthiest, bloodiest places" he'd ever seen. Offended by the smell of discarded seal carcasses rotting beneath the sun, the visitors stood at a cautious distance to watch the skinning procedure and take photographs. The Indians had stretched out seal meat to dry on the beach to provide for their long winter, and the animals' oil had been stored in square wooden boxes, which the Indians had carved and decorated to hold the precious liquid. Covered with the blubber they extracted from the seals, the women worked hard at their tasks, fascinating Grinnell with their skills. He carefully watched the entire process, noting their efficient tools and deft handling of the carcasses, but photographs to document the procedure were hard to obtain. The Indian women scowlingly turned away from the cameras aimed toward them. John Burroughs observed the photographers "watching and waiting and maneuvering to get a good shot." To the naturalist, photography could never be an art. The mechanized process, he thought, did not "involve the imagination." But Edward Curtis knew better. He sought an artistic style that would convey a sense of the mysterious and mystical in Indian life. A good photographer, he was learning, had to establish a friendship with his subject, particularly if he sought a portrayal of the American natives. Curtis wandered down from the Malaspina Glacier to take prints of the Indian women at their work.

John Muir could not bring himself to approach the village. While the stench of decaying seal carcasses was bad enough, he could still hear the eerie noises made by the hunted seals in the night, "barking or half-howling in a strange, earnest voice." The naturalist would content himself instead with exploring the glaciers and inspecting the flowers blooming around the bay. Had he ventured near the seal hunters' camp, Muir would undoubtedly have been alarmed to see five sea otter pelts that the Indians offered for sale to his colleagues. The sea otter was a nearly extinct animal, and its skin brought high returns for the native hunters. The Indians took full advantage of the rarity of the animal, refusing to budge from their prices of three to five hundred dollars apiece for the furs. Despite the high price and the heedless hunting of the endangered animals, the Indians found a buyer in Harriman, who selected the richest, finest pelt they had to offer.

Seal hunting looked exciting to Merriam and Curtis, and the two decided to participate in the Indians' ritual. They paddled out near the icebergs where the seals splashed about in the water, but killing the animals proved to be far more difficult than it looked from shore. Curtis finally managed to shoot one, but the wounded seal slipped away into the icy water. If seal hunting lost its appeal, Yakutat Bay also abounded with cod and salmon, as well as numerous birds for the ornithologists' rifles. Seeing salmon darting through the clear water off shore, Averell and Trudeau, eager for hunting of any sort, grabbed their rifles and dashed down to the beach to shoot the fish as they swam past. Mrs. Averell found a line and pole more to her liking. In genteel fashion, she passed the leisurely days at Yakutat Bay fishing for cod over the side of the *Elder*. One fish she caught yielded an unexpected find for the scientific collectors: two long parasites clung to the side of her catch, and Dellenbaugh quickly rescued the strange-looking specimens and gave one each to the delighted Ritter and Coe—the "worm men," as the artist called them.

The long hours of daylight and the changes in time zones disoriented the travelers on board the *Elder*, and the days spent in the land of the Indians drifted together for the expeditioners. At a small general store in an Indian village, members of the Harriman party struck up a conversation with the white storekeeper, who revealed that the Harriman women were the first white women he had seen in over a year. The information shocked Keeler. He was stunned at how far from "civilization" he had wandered. The old storekeeper had come to Alaska seeking gold ten years earlier, and on occasion he still took his battered old prospector's pan out to the nearby streams. Pulling out a folded newspaper, the storekeeper carefully unwrapped the sandy mineral it held and poured it into a pan of shallow water so that his curious customers could see the golden colors it held. The wet glittering mass intrigued Harriman, and he promised the prospector that he would bring a mineral expert, Walter Devereux, to assess the possible treasure.

Devereux, however, was camped down the bay near Merriam's party, seeking his own treasure—a bear. Days of hunting had produced only tantalizing traces of the animal, and after several days in camp with Morris and Grinnell, he was ready to return to the *Elder*. After three days of slapping mosquitoes and with no bear, the men "rejoiced to see smoke from the steamer in the distance, far away over the drift ice." Once again, the great ship could not force its way through the dense ice and Harriman lowered two boats, with himself at the helm of one, to rescue the stranded parties. Picking his way among sharp ice floes, the railroad tycoon paddled several miles between the icebergs and the shoreline to reach the hunters and scientists. By that time, rescuing his guests had become a familiar job for Harriman. Only the day before he had rowed near the Indian village through a high pounding surf and dangerously jagged ice to reach Gilbert, Gannett, Muir, and Kearney, who waited on shore. Kearney, the youngest member of the small party, was more than ready to enjoy the comforts on board the ship once again. Blisters covered the palms of his hands from the constant rowing among the icebergs of the bay; Gilbert, elected the leader of the party, had allowed the others "little rest from the oars." But Muir's lively storytelling at night before the fire had been entertaining and, years later, Kearney still recalled the sound of Muir's thick Scottish

accent as the famous naturalist repeated tales of the unpro-
voked beatings his strict Scottish father had given him as a
small boy, an experience apparently never far from his mind.

But for some members of the expedition, the time spent at
Yakutat Bay passed too quickly. The idyllic bay offered
Charles Keeler his first camping trip, and the poet pitched a
tent alongside Ridgway, Ritter, and Saunders near a small inlet on the bay shore. Despite pesky mosquitoes and drizzling
rain on occasion, Keeler found camping a refreshing respite.
He sat by the campfire at night, listening to the birdcalls and
idly chatting with his companions. He felt proud of himself
for withstanding the cold and rain of camping with no ill effects; the trip, he felt, had already improved his health.

On a glacier high above Keeler's camp, Gannett and Gilbert
put their time to good use. They were carefully surveying the
glacier when the *Elder* whistled for them to return. The two
scientists, however, refused to return: there was not enough
time, they thought stubbornly, to finish the task at hand. The
calls of the *Elder* finally won out, however, and after keeping
the pilot waiting impatiently for five hours, Gannett and Gilbert deserted their work and disappointedly headed back to
shore to be picked up.

By June 23, all the parties had regrouped back on the *Elder*.
After Albert Fisher checked his mousetraps one final time and
Harriman hurriedly bought three Indian canoes at the last
minute, the ship steamed out of the bay as bad weather brewed
overhead. Prince William Sound, farther up the Alaska coast,
was next on the expedition's schedule. As the ship steamed
along, a majestic albatross, its wings spread against the dark
clouds, followed it, escorting the expedition along its way. The
scientists clustered at the rail of the deck, watching the graceful bird dip and soar as it kept pace with the ship. To John
Burroughs, undoubtedly familiar with Coleridge's "The Rime
of the Ancient Mariner," the bird seemed somewhat ominous.
Entranced by the effortless flight of the albatross, Burroughs
imagined that it was "like the spirit of the deep taking visible

form and seeking to weave some spell upon us or lure us away to destruction." The land of Indians and silent glaciers had worked its strange magic on Burroughs' lively imagination.

But at nightfall the romantic notions left Burroughs' head as he listened to the scientists assess the discoveries at Yakutat Bay. In the lecture hall, Merriam, Kearney, Ritter, and Gannett talked earnestly of the area's geography and its plant and animal life. Later, the discussion turned to less academic subjects in the casual smoking room. Most of the men listened rather uncomfortably to a political diatribe given by one of their colleagues. Animosities from the Civil War and Reconstruction still lived on in the orator, who berated the North for feeding "upon the South like vultures feasting on carrion." Politicians, he argued, "would have their hands in the pockets of naked savages" were they only given the chance. His listeners smiled politely, but at Orca, the expedition's next stop, the party would discover that it was the entrepreneurs of a large salmon cannery—who made a lively profit from carelessly exploiting both Alaskan natural resources and Chinese laborers—who had their hands in the "pockets of savages."

CHAPTER EIGHT

"THE MAP'S VOID SPACES"*

The long westward voyage to Prince William Sound lasted a full day and night, and the scientists passed the time telling tales of their adventures. Men like Muir and Brewer, old veterans of Alaska, carried an inexhaustible supply of stories in their heads. At Nuchek, one of the expeditioners pointed out the Russian trading post where years ago a thief had robbed a fur trader of thirty thousand dollars' worth of skins only the day before the steamer arrived to cart them off. But a loud blast from the ship's bell interrupted the storytelling competitions: Captain Doran had ordered a fire drill. Suddenly the deck came alive with frantic crewhands, who grappled with the great water hose as the bell clanged, while Doran barked instructions to lower all boats to the water. The *Elder*'s passengers had never seen such frenetic activity on board. Two loud blasts on the whistle called all the expeditioners to assemble on deck, dutifully clad in their life preservers. Doran passed among his passengers, efficiently inspecting each preserver and pointing out its proper usage before he dismissed the group.

*From "The Call of the Wild," in *The Spell of the Yukon* by Robert Service.

Removing the bulky, awkward preservers, the party found the boat easing into Prince William Sound, where the ship finally came to rest off shore at the small town of Orca.

The idea of stretching their legs on solid ground appealed to most of the expeditioners and several boats were soon splashing through the water toward town. Dominated by the Pacific Steam Whaling Company's salmon cannery, Orca boasted three buildings and the foul odor of rotting fish. For miles along the coastline, discarded salmon heads and fins littered the ocean, lending the water an oily look. About two hundred Chinese laborers worked long hours in the cannery, and Muir sadly shook his head at the sight of the men brought up from San Francisco to work for low wages. "Men in this business," he wrote, "are themselves canned." Muir quickly left the scene and visited the only store at Orca, where he was flattered to discover that the owner had read his book *Mountains of California*.

Dellenbaugh had set up his easel on the ship's deck to sketch the mountains in the distance. Pen in hand, he concentrated intently on the drawing until a passing scientist reminded the artist that, despite the full light, it was already after ten o'clock. Startled, Dellenbaugh looked around him—in New York, he thought, the light would indicate the time was late afternoon. Under the balmy breeze and clear sky, he continued drawing until midnight, still able to see his work clearly. Meanwhile Edward Curtis, high on the mountain above Orca, was photographing as late as eleven p.m. Like Dellenbaugh, he focused on the immensities of the landscape. Neither the painter nor the photographer chose to depict the grisly realities of the cannery or the depressing realities of the town itself. Instead, they concentrated upon producing luminist dreamscapes.

Burroughs, on the other hand, found the town alive and fascinating at the late hour, and spent the evening visiting with several gold miners he encountered on the beach. The men had hiked back from their futile prospecting sites on the nearby Copper River and were waiting for a steamer to arrive that night to take them back to their homes in the States. Burroughs listened sympathetically to the tragic stories of suffering and disillusionment the miners told. Several thousand men had hurried to the Copper River during the past year seeking the gold rumored to be easily obtained in its waters. But of the group camped amid the salmon remains on the Orca beach, none had found gold, though all had felt fortunate to have survived the hardships along the gold route. With a compassionate sensibility akin to that of his friend Walt Whitman, Burroughs wrote of the spirit of these common men: "They were from the East and from the West, lumbermen from Maine and Pennsylvania and old miners from California and Colorado. They were a sturdy, sober-looking set of men . . . no nonsense about them." The previous spring, nearby Port Valdez had been the site of an outbreak of scurvy, which took the lives of many of the gold miners who sought an overland route to the Klondike from Valdez. The men camped forlornly at Orca were the survivors. The disease had crippled one older man in the group, and his son, Burroughs noted, watched tenderly after him. The young man confided to Burroughs that most of the gold miners were scurrilous "Eskimo humpers," whose morals would appall him. Stranded with no money, the miners waited for the steamer and hoped that its captain would be generous enough to take them back to San Francisco. These, Burroughs thought, were "homely, slow, deliberate men but evidently made of the real stuff." He was relieved to see the miners boarding the steamer the next day, relinquishing their dreams of gold for the relative comforts of their homes.

After the ship had carried the ragged miners away, Dellenbaugh befriended one gold rusher who wandered up to watch him paint on the beach. The young prospector had refused to give up the search for instant wealth, and passed the time at Orca waiting for money from home "to try his luck once more" in the Klondike. The lure of gold had become a virtual obsession. The miner told Dellenbaugh of a friend who, while dying painfully of scurvy in the Klondike, spent his last days mending his socks so that he could prospect again when he regained his health. The sound of the *Elder*'s whistle calling its passengers curtailed further stories, and all returned to find that two new guests had boarded the ship at Orca: Harriman had invited officials of the Pacific Steam Whaling Company and the Alaska Commercial Company, Captain Omar I. Humphrey and M. L. Washburn, to sail the coast with his distinguished party.

Under clear blue skies that were unusual for an Alaskan summer day, the *Elder* steamed onward through Prince William Sound, discovered long ago by Captain Cook. The expedition cruised leisurely through the sound, depositing various parties of scientists at the sites they pointed out. At Columbia Glacier, Gilbert decided to remain behind for three days, accompanied by Palache and Coville. From the placid Prince William Sound, mountain ranges rich with the golden glow of soft light stretched in every direction as the ship puffed forward and finally came to anchorage in Golofnin Bay. There, a large group went ashore to explore for several hours, and Fuertes set out in search of birds. Harriman, accompanied by "Indian Jim," rambled over the mountains, trusting the Indian to choose their route. At midnight, they returned hungry but exhilarated after their long climb, though still no bears had been spotted.

While Harriman was off climbing, a visitor approached the *Elder* in his own small boat. The newcomer ran a fox farm on one of the many islands in the Sound, and he hoped the blue-hued pelts would bring a large profit when it was time to skin the animals. Perhaps, he suggested, the expedition might enjoy a tour of his enterprise during their stay in the area. The entrepreneur had originally bought two pairs of blue foxes from the Alaska Commercial Company and now his farm contained forty, each of whose skins, he estimated, could bring from between fifteen to thirty dollars. Entrepreneurship, it seemed, thoroughly pervaded Alaska's coasts, and even the smallest islands held men with business schemes. This had been Alaska's heritage from the very beginning, when the first Russian seal hunter descended upon its shores. A pristine wilderness that seemingly beckoned the man of nature, like John Muir, it had

seen few such men. Rather it had entertained for the most part
ruthless exploiters from all nations—Russia, Britain, the United
States, and even France. Alaska had always been a bonanza, a
colony, a "profit point," and rarely the romantic park that many
of the expedition members seemed to think it was.

Characteristically, most of the party were more interested
in exploring the shore than in touring the fox farm: the Harriman women piled into a launch along with a "pleasure party"
to visit the coast under the pale light of the long evening. A
party of scientists had pitched camp near a stream where a log
cabin stood, and the group made their way toward the spot.
Seeing the company approach, a Norwegian prospector bus-
tled out of the log cabin, sweeping his hat from his head at
the sight of the women and gallantly bowing to them. In the
year since he had settled at the cabin, the Harriman women
were the first ladies to visit his home and it was, he assured
them, quite an honor. Copper mining yielded the old Norwegian a more than adequate living; indeed, he planned to
travel to the Paris Exposition in 1900. Perhaps more than furs
and gold, the region's copper promised a steady income for
those who relentlessly sought their fortunes in the Arctic wilderness.

In the northeast corner of Prince William Sound, Port Wells
extended a narrow channel into the icy land and lured the
scientists with its reputation of immense glaciers and unknown
life. The next day, Burroughs shivered in the "great ice chest"
of the glacier-surrounded Port Wells. The scientists, less sensitive to the cold, excitedly observed the glorious region, which
few whites—not even the well-traveled Muir—had explored.
Gifford and Dellenbaugh set their palettes in a launch and
rowed to shore to paint the wild scenery; boats of scientists
followed, eager to see what the strange coast would yield for
their study. Gannett and Muir were anxious to use their little
time in that location charting the unknown glaciers, and, since
Harriman wanted to take his children on an outing, the two
glaciologists surmised that there might be time to complete the
surveying. It seemed, however, that they had barely begun

when the ship's whistle signaled for their return. The frequent conflicts of schedules among the scientists and hunters often resulted in frustrations and incompleted work.

But a sudden turn of events soon made the two scientists glad they had returned on board. When the *Elder* steamed into sight of Barry Glacier, Captain Doran announced that U.S. Coastal Survey maps indicated that navigable waters ended at the glacier. But as the ship edged closer to the great wall of ice, a thin inlet of water became visible, stretching out beyond the Barry. Excitement over the discovery spread quickly throughout the *Elder*, and passengers poured on deck to catch a glimpse of the sparkling, unknown fjord. "We shall discover a new Northwest Passage!" proclaimed the ever-ambitious Harriman, ready to charge the great ship through the narrow passage. Though Captain Doran hesitated at the possible danger, Harriman reassured him with an acceptance of full responsibility for the risk. However, their guest Captain Humphrey, who had already proudly discussed his expert knowledge of the area, dismissed the new "find" with a harumph. There was certainly no point, he grumbled, in charging recklessly into "every little fish pond." The unseen rocks beneath these waters, he knew, could be extremely dangerous to a ship the size of the *Elder*.

Harriman's entrepreneurial leaps of speculation had consistently landed him in safely powerful positions in the business world, and he seemed to take for granted the success of his brash actions in the physical world as well. Later, John Muir would write of his host's unwavering determination: "Nothing in his way could daunt him or abate one jot the vigor of his progress toward his aims, no matter what—going ashore through heavy breakers, sailing uncharted fiords, pursuing bears, etc." So it was that, against Doran's and Humphrey's advice, the tycoon stubbornly ordered the ship to proceed "at full speed, rocks or no rocks." His gamble, as usual, paid off, and Humphrey's "fish pond" opened up into yet another long fjord, curving away past the front of the Barry Glacier and revealing 'a snowy iceland of countless glaciers never before seen by whites. Like giant polar-bear skins, marveled Daniel Elliot, the glaciers stretched far back into the wilderness. It was the Harriman Expedition's major geographical discovery.

John Muir fairly itched with excitement to hike about these fine glaciers. Along with Gannett, Gilbert, Inverarity, Kelly, Palache, Coville, and "Indian Jim," he pitched camp on the shore to explore the rugged terrain. True to Humphrey's fears, one of the *Elder*'s propellers had snapped from an impact with a hidden rock, and since the ship would have to return to Orca and dock for repairs for two days before returning, the scientists were assured that on this occasion, at least, they would have time to complete their surveying. A half moon hung low in the sky, rising over the snow-covered mountains, and the campers pitied their colleagues on board the ship for missing the peaceful glory of time spent in the silence of the wilderness. Here in Prince William Sound the expedition would leave a lasting monument to itself. Later maps would designate "Harriman Fjord," extending past the Barry Glacier, and

The Harriman Fjord—The Serpentine Glacier, by E. S. Curtis.

"Harriman Glacier," the largest of the fjord's many glaciers. Though he had found no new Northwest Passage, Harriman's impulse to explore the dangerous waters had yielded an important new geological find.

After a celebratory evening dinner of Welsh rarebit, Dellenbaugh, Gifford, and Brewer stood at the rail of the deck long after their colleagues had retired. The three men were curious about the intricate maneuvers the ship would need to make to pass through the narrow inlet at Barry Glacier. Hugging the side of Barry Glacier as close as safety would allow, the *Elder* edged slowly past the narrow opening back into the larger area of Port Wells, crunching its iron hull ominously on hidden ice but pulling through safely. Harriman relaxed below deck with Morris and Trudeau, coolly playing his favorite board game— Crokinole. Over and over he snapped round discs on the board in an attempt to land the pieces in a center ring, while the ship's pilot tried his best to guide the *Elder* out of the fjord.

All afternoon, the *Elder*, minus one propeller, crept back to Orca, followed by a playful school of more than a hundred dolphins, who frolicked in the wake of the ship. Late in the evening, the expedition reached the town, playing its graphophone at full volume to entertain the crowd of gold miners watching from the pier.

Under heavy, threatening clouds, the expedition members optimistically proposed a picnic the next morning to pass some time while waiting for the repair of the propeller, and two launches were hauled off the ship to bear the Harriman family and several others to a bay nearby. In the Harriman custom, it was no ordinary picnic: a pilot, a cook, and a waiter accompanied the group to the site, where a tent and fire already awaited them. Among those remaining behind was Fuertes, who had decided to try his hand at paddling one of the Indian dugout canoes. After he persuaded Trudeau to join him, the two climbed awkwardly into the small boat, folding their legs into the half-squat demanded by the canoe.

Brewer and Burroughs took the opportunity to tour the salmon cannery they had visited before. One's taste for the

fish, Burroughs soon found, disappeared at the sight of the
workers standing in the midst of discarded salmon parts.
Working "like jugglers," thought Burroughs, the Chinese
flashed their sharp knives about the fish so quickly one could
barely see the procedure before the sliced fish were tossed aside
into great barrels.

The use of Chinese labor, like the stripping of the Alaskan
streams of their salmon, had been a continuing source of con-
tention between the natives and the white corporate interests.
By 1899 at least fifty-five salmon canneries along the Alaska
coast busily hauled the fish, as well as the profits, out of the
region. Ignoring the economic needs of the Alaskan natives,
the canneries found migrant labor cheaper, and they transported the Chinese up the coast from San Francisco to work
for pitiful wages. The workers frequently received their meager paychecks only after they had been deposited back in San
Francisco when the seasonal work ended. The combination of
cheap labor and abundant natural resources produced a thriving industry, whose business had increased over the past two
decades at a phenomenal rate. When Burroughs and Brewer
toured the Pacific Coast Whaling Company's cannery, salmon
enterprises represented the biggest industry in Alaska and several of the largest companies had banded together in a "combine," the Alaska Packers' Association. At Orca, though, the
Pacific Steam Whaling Company had refused to join the trust
and a lively competition for markets ensued. But a graver concern than the combine loomed on the horizon—the greed of
the canneries threatened to diminish the supply of salmon,
which had once been abundant.

Three years earlier, the government had passed legislation
that prohibited salmon fishing in specific areas, but conservation measures in the remote regions had been impossible to
enforce. The native Americans suffered as their food supply
poured into the cans shipped by the companies to the States.
One Wrangell Indian pleaded in broken English for the needs
of his people: "By and by [white people] began to build canneries, and take the creeks away from us, where they make salmon

and when we told them these creeks belonged to us, they would not pay any attention." In an era when railroads like Harriman's Union Pacific thundered across the country, increasing the market for such Alaskan resources, the Indians' simple lives could not withstand the economic pressures of industrialization. It was perhaps with an understanding of the natives' imminent cultural decline that Harriman recorded the Indians' songs and voices. But the railroad tycoon had a solution for the problems of the natives: the corporate interests that stripped the natives of their culture and economic livelihood could use Indian and Eskimo labor in the lumbering businesses, the canneries, the mines, and the fur-trapping enterprises that were rapidly changing the face of the land.

Harriman's photographer Edward Curtis, however, would not have shared the businessman's eagerness to integrate the Indian into a foreign labor force. Instead, preserving the culture of the beleaguered race would become Curtis' lifelong task. During the long days on the expedition his friendship with George Bird Grinnell grew into a strong mutual respect, and that evening, after the *Elder* had been furnished with a new propeller, Curtis heard Grinnell lecture on the Blackfeet Indians, among whom the ethnologist had spent many days. The stories Grinnell told touched Curtis; it became his dream to record the customs and, indeed, the very faces of these people for posterity.

On the opposite end of Port Wells, Gilbert, Gannett, Muir, and Palache carefully recorded the measurements of the new glaciers in Harriman Fjord for government logs. While Grinnell entertained an audience in the *Elder*'s lecture hall with his Indian tales, the ship steamed across the sound to pick up the party of surveyors. At six in the morning, the ship's whistle pierced the cold air of Harriman Fjord and the party of campers woke with a start. They gathered their instruments, maps, and camping equipment, and boarded the ship, ready for a hearty breakfast. Muir felt ecstatic after the days in the awesome isolated fjord, and he hoped that the *Elder* would stop at the glaciers in "Icy Bay" as it steamed toward Cook Inlet in search of better bear-hunting grounds. But Harriman shook his head; there was "no more ice time," and he had the ship stop instead at a copper mine on LaTouche Island that he and Devereux wanted to inspect.

While Harriman, Devereux, and a couple of other men briefly surveyed the copper operation, several passengers on board the *Elder* watched anxiously as an enormous killer whale, perhaps curious about the large vessel, approached the anchored ship. An incredulous Merriam estimated the creature to be at least eighty feet long. Later, as the *Elder* steamed away from the island and out of Port Wells, even more whales joined the first. Three of the great beasts, "standing up like huge stumps above the water," threw their enormous dorsal fins high into the air before they dived beneath the water. They seemed to tease the ship as it puffed away past Orca and up to Cook Inlet. En route, during a stop at Saldovia Point, a Russian native advised Harriman of the game and the pleasant climate at Kodiak Island. Convinced that Kodiak offered him a better chance to bag a bear, Harriman ordered the *Elder* to turn its course southwestward and head for the large green island off the Alaskan Peninsula.

CHAPTER NINE

"DOG DIRTY AND LOADED FOR BEAR"*

Volcanic smoke wafted lazily from Mount Iliamna across Cook Inlet as the *Elder* changed its course at Harriman's command and headed for better hunting grounds on Kodiak Island. The ship sailed all day beneath clear blue skies that made the coastal waters shimmer and sparkle, but by the time the expedition reached Kukak Bay on the Alaska Peninsula, where a group of determined botanists and ornithologists had decided to camp, darkness had settled in and the group was forced to lower their launch in unfamiliar waters made more hazardous by night. John Burroughs worried as he heard their boat splashing away through the darkness, headed toward a shoreline that was several hours of hard rowing away. The days of collecting at Kukak Bay, the scientists hoped, would prove more rewarding than following Harriman's hunting excursion to Kodiak Island.

A small band of hunters also had a separate destination in mind. Led by Dr. Morris, Grinnell, Devereux, Averell, and Trudeau chose a campsite at Uyak Bay on Kodiak Island, far

*From "The Shooting of Dan McGrew," in *The Spell of the Yukon* by Robert Service.

from Harriman's eventual anchorage point on the opposite side of the island. The *Elder* stopped at Uyak Bay early in the morning to let the party off, and while the others watched from deck, the hunters struck out on their quest for bear, taking all the packers with them. The men set off across Kodiak's green, rolling hills, a gentle contrast to the rugged, stark majesty of the Prince William Sound mountains and glaciers. To Muir, the hunters seemed incongruous with the idyllic island, and he watched them bemusedly as the hopeful band marched off, "gun laden as for war."

One hundred miles away lay the village of Kodiak, or St. Paul, a small settlement where the *Elder* would anchor for several days while Harriman sought his own camping grounds. As the ship steamed between the coasts of Kodiak and Afognak Islands, Harriman's new guest, M. L. Washburn, the head of the North American Commercial Company at Orca, frowned at the sight of the other island. Here, he told his companions, was a clear case of governmental folly. Regulations forbidding hunting on the island were a "farce" since the government never even attempted to enforce its laws. Besides, Washburn continued, if the Afognak reserve was meant to protect the endangered sea otter, the government had surely chosen the wrong site. A far better reserve would be the outlying islands like the Shumagins, where the otters actually congregated. As it was, the regulations merely spoiled the game hunting on Afognak without protecting the rare animals of the region.

Though Afognak was off limits, Harriman hoped that Kodiak held similar game and, as the *Elder* pulled into the narrow dock at St. Paul, he wasted no time readying equipment for his own hunting trip. John Burroughs, however, was too enthralled with the sleepy village, nestled against a wall of green hills, to think of his host's business on the island. Tired of the craggy, intimidating peaks of Glacier Bay and Yakutat, the naturalist reveled in the "placid" and "pastoral" meadows of the island. Even the village itself had a rural feel to it. Only one store served the community, and its great weatherbeaten sign announcing the "Chicago Store" amused the Harriman visitors. The scattered cottages boasted small gardens, while chickens and cows roamed freely about the town. Farming was difficult in the climate and soil on the island, and the U.S. government had just begun experimentation on Kodiak to determine what crops might prove profitable for the region.

Few Americans lived among the population of several hundred Russians and Indians, but among the Americans at St. Paul was an old acquaintance of Dall, Gannett, and Merriam; the three men sought him out and he entertained them with bear-hunting tales, relishing his rare visitors. The old hunter had sent Merriam, who was writing the definitive treatise on the "old bruin," enormous bear skulls from the island. His wife, he said proudly, had killed the largest bear when the beast had surprised her in the woods.

Merriam hurried back to Harriman brimming with new enthusiasm for the possibilities of hunting bear on the island. Together with his oldest daughter and his niece, Harriman instantly set out with Merriam in a launch to set up a hunting camp at Eagle Bay, eight miles away. As Harriman steamed away, most of the expeditioners explored the village and its surroundings. Streets of grass wound about lazily among the modest cabins; since there were no horses or wagons on the island, transportation was all by foot. The tall spire of a Russian church rose from a hill overlooking the town, and its chimes filled the air as Dellenbaugh and Gannett began climbing the small mountain behind the town. Spongy moss padded the incline, and wild geraniums and forget-me-nots bloomed freely on the hillsides in patches of red and blue. Burroughs and Muir already stood atop the mountain when the others reached on the summit, and together the four men took in the sweeping view below, unencumbered by trees. The sight nearly took Muir's breath away. "No green mountains and hills of any country," he wrote, "can surpass these of the Alaska Peninsula and the Aleutian Islands, where no trees grow . . ." For Burroughs, Kodiak Island was paradise after the "ice chests" of the glaciers and he sat peacefully on a knoll, his hands filled with wildflowers.

The next morning, John Muir woke to a "calm, misty" Sunday that promised another tranquil day. But, to his annoyance, it seemed that virtually all his companions had plans for hunting, and were toting their rifles and "sauntering as if it were the best day for the ruthless business." Along with Trelease and Gifford, Muir deserted the cocky hunters and headed again for the lush greenery of the foothills east of St. Paul. Dellenbaugh set out to find a site for his sketching and, after wandering through the grassy footpaths, he settled into a spot in a rundown section of town where Indians and poor Russians lived. Garbage, he noticed, had been thrown haphazardly outside the house, and the odor of fish carcasses wafted over on the breeze. Nevertheless, he thought democratically, the people themselves were certainly as "pleasant and cordial" as his elite company in Sitka had been. St. Paul, on the whole, was as "charming" as Yakutat Village.

At the Harriman camp eight miles away, the visiting hunters busily came and went. Dall, Elliot, Coville, and Palache showed up in a steam launch, accompanying Mrs. Harriman, her daughter Cornelia, and Dorothea Draper. But they missed Harriman, who had already struck out on his bear hunt, accompanied by Captain Kelly and a Russian villager named Stepan Kandarkof for a two-day expedition. Here, the railroad tycoon hoped, would be the climax of his vacation. Meanwhile, Coville and Palache, eager to entertain their young female company, took the four girls in a boat upstream. Several miles upriver Palache spied a young bald eagle in a nest in a cottonwood tree and impressed his company by climbing up the branch and seizing the bird for an addition to the scientific collection. Soft, dense down covered the bird "like beaver fur," and back at camp, Merriam excitedly examined the eagle. It would make a fine contribution, he declared, to the specimens they would ship back to Washington.

By the following day, only Merriam remained at his host's

hunting camp and the scientist kept a solitary vigil for Harriman. The old biologist hoped for the best for Harriman's hunting efforts, which had failed so often in the past month. Merriam knew how important the trophy of a bearskin was to Harriman; after all, it was, ultimately, the lure of bears that accounted for his own and the other savants' presence in Alaska. When Harriman arrived gleefully back in camp around nine in the evening, Merriam sensed from his expression that this time the hunting had been good. The railroad tycoon proudly announced that at last he had attained his goal. A dead bear lay several miles inward on the island, a victim of Harriman's rifle. Captain Kelly, too, arrived in camp feeling somewhat victorious. Harriman's female bear had been accompanied by her small cub, and Kelly had shot the infant bear. The bear, wrote Merriam later, was "a real big genuine Kodiak Bear," and he shared Harriman's delight in the successful hunt. It was not, however, a particularly grand episode in wilderness hunting. The bear had been driven down into a narrow gorge, where Harriman awaited his prey with a powerful rifle. His company had taken great care to ensure their host's

Bear Hunters at Kodiak Island (Harriman in center).

The Bear, by E. H. Harriman.

Stepan Kandarkof—Russian Bear Hunter, by E. S. Curtis.

safety, and Kincaid wrote bitingly that "lest the bear behave in an unpleasant manner a group of hunters were grouped about [Harriman] with enough firepower to tear the bear to pieces." But, fortunately for the bear and for the quality of Harriman's new trophy, one shot by the tycoon had killed the creature. Feeling a new camaraderie, Merriam and Harriman rowed with Kelly and the Russian back to the *Elder* to break the good news to the rest of the expedition. It would be up to

Fourth of July at Kodiak, Alaska, by F. S. Dellenbaugh.

volunteers from the expedition to haul the heavy beasts back to the *Elder*. On hearing the location of the two dead bears, Fuertes and Cole, the taxidermist, set out at three in the morning to skin the animals and carry the trophies back to the ship. The next day's Fourth of July festivities, they thought, would not be complete without the evidence of Harriman's hunting prowess—if not the American dream, it had surely been Harriman's Alaskan dream.

On the morning of July 4 crew members of the *Elder* busily stuffed old rags into the small brass cannon planted on the hurricane deck. When fired, the rags made a gala explosion to begin the day's festivities, and for a solid hour, the officers shot the little cannon off into the air. Two miles away at Wood Island, where the day before Mrs. Harriman had led a picnic, a ball game between the Kodiak and Wood Islanders was starting and the expeditioners steamed over in their launches to watch the natives compete. Later in the afternoon, the *Elder*'s deck would be the site of the expedition's own program of entertainment. The Committee on Music and Entertainment—Dellenbaugh, Fernow, Fuertes, Cornelia Harriman, and Dorothea Draper—had faced their first real duties, and had lined up a series of recitations, speeches, dances, and music to commemorate the day. Little Carol Harriman was determined to be part of the program, and, as Dellenbaugh's boat floated toward the Aleut village on Wood Island in the morning, the artist composed a poem for Harriman's youngest daughter to recite. With the poem written, the agenda for the afternoon was completed and Dellenbaugh managed to relax at the Wood Island ball games.

Back on the ship after noon, the graphophone burst out with John Phillip Sousa's "Stars and Stripes Forever," and Brewer gave an extemporaneous address. The United States, asserted Brewer, was "the first to fight for humanity," and in the recent Spanish-American War the country had set a worldwide example of its dedication to freedom. But Charles Keeler had written a poem that launched into zealous denunciation of the United States' current policy of intervention in Manila. Taking the stage, he read his "Fourth of July Ode" to the assembly:

> Is this the wilderness—these greensward hills,
> These wastes of lupin, wildflowers and of rose,
> These slopes of heather, by the mountain rills,
> O'erhung by skies of gold through day's slow close,
> Where one long lotus-dream obscures all human woes.
>
> Here sing the birds on hill and in the glade,
> The warblers flash afield like waifs of gold,
> The thrushes chant their vespers in the shade,
> The varied robin's pipe afar is rolled.
> And in the little church the bells are clanged and tolled.
>
> And we who tarry here this festal day
> Still see the flag of home wave proud on high,

Still find a welcome on our seaward way.
For where the flag waves, home and friends are nigh,
The eagle flaps his wings and makes exultant cry.

His cry is liberty, as heaven's high dome
He scales on peerless wing; and we so kind
Should back our answer as we westward roam,
Trusting our voices to the heedless wind,
That roams the misty sea, a pilgrim lost and blind.

Call ye this liberty, where law's strong hand
In nerveless palsy falters over wrong?
Sing ye of freedom in a lawless land?
The very winds shall mock your idle song
And in a wail each syllable of pain prolong.

Ye who have failed to rule a wilderness,
Now preach of liberty in tropic seas,
Forsooth our sway the Orient hordes shall bless,
While politicians fatten at their ease;
O Lord, must our dear sons be slain, such men to please?

O teach us in the wilderness thy ways,
And by the mountains let thy law be sung;
No thing of earth shall stand, which disobeys,
Thy bidding, every clod shall find a tongue.
And liberty by bell innumerous shall be rung.

A dismayed Dellenbaugh watched the crowd stir uncomfortably. "Decidedly out of place," he said, frowning, surprised by the entrance of political controversy into his program. But young Carol Harriman relieved the tension when she haltingly recited the poem Dellenbaugh had prepared for her. She lifted the spirits of the audience with a patriotic appeal that Dellenbaugh titled "Our Banner":

> From tropic seas to midnight days
> Our banner flies, a promise clear;
> From East to West, the sun always,
> Follows the flag that knows no fear.

Our navy's guns awake the world
And crush the tyrant's evil power.
Oppression's flag is quickly furled
Where Columbia rises to the hour.

Thus stands our country with the great,
Where other lands have often stood;
Yet hard! To hold this envious place,
As well as great, we must be good.

Then fling our emblem far and near,
In every clime, and every sun;
But let us all remember here,
That greatness, goodness are as one.

Applause burst out when she proudly concluded the last verse. As her ovation faded, Gifford tucked his fiddle beneath his chin and gaily sawed away. His colleagues Fernow and Ritter cast dignity to the wind and danced a light-footed clog—followed, rather incongruously, by the singing of the Doxology en masse. Dellenbaugh sighed with relief as his part of the program ended and the traditional local canoe race began.

Long a custom among the islands, the races matched dugout canoes with bidarkas, and rowboats against other boats. The races themselves were segregated; Indians raced against Indians and Aleuts, while whites raced only against others of their race. Disruptions plagued the day's sport from the start. When two Kodiak boats set out to begin the first competition, the first boat sprang a leak and limped back to shore, ending the race. The second race, for rowboats, was inadvertently joined by Yakutat Indians in a bidarka canoe. The Indians took off eagerly at the sound of the starter, paddling with all their strength across the water, oblivious to the calls from shore that they were in the wrong race. And, in the next race, the onlookers laughed heartily on the pier when a team of Yakutat Indians suffered the embarrassing misfortune of capsizing their canoe just as they reached the finish line. Harriman himself ran a close race in a four-man canoe against a naphtha launch, while his excited guests cheered him on. Despite the mishaps that plagued the races, the Harriman expeditioners left the competition in high spirits. When Fuertes and Cole finally paddled up to the *Elder* carting the two dead bears, the entire party sent up a loud cheer.

All, that is, but Muir. The bears lying dead in the little rowboat were, to the naturalist, "mother and child," and he gazed on all the celebration scornfully. Merriam's label of "genuine Kodiak bear," though accurate, failed to indicate the size of the animal, which was actually *small* for the species of bear. Glancing at the female animal's hind foot, he estimated it only about eight inches. Muir sought refuge from the celebration by visiting with a grizzled mountaineer who, having heard that the famous John Muir was on board the large ship docked in the harbor, made his way to meet a kindred spirit. The old man shared Muir's affinity for exploring the glaciers and mountains by himself, alone in the vast, silent reaches of nature. Although life in his small cabin on Cook Inlet sometimes grew lonely during the long Alaskan winters, his solitary adventures compensated for the lack of company. Muir, delighted that the isolated mountaineer "had heard of [him]," traded stories of travel and exploration with his new friend for hours.

The day wound down pleasantly and Harriman, now that his bear trophy had been won, much to the admiration of the expeditioners, looked ahead to a daring sidetrip across the Bering Strait to the Siberian shore. When the ship began stoking its fires to start out of the Kodiak dock at midnight, Keeler, perhaps nervous about the upcoming voyage to the rocky Russian coast, wrote to reassure his wife about the trip. "Everything," he penned in a neatly curvaceous style, "is done in the most luxurious manner possible." But still, he silently dreaded leaving Kodiak for the rough seas ahead. The island, he mused, was "the most beautiful place we have encountered in Alaska." Keeler did not have long to fret before the *Elder* struck an obstacle not far from St. Paul harbor; a sandbar grounded the ship until the following morning's tide lifted it off its temporary landing. The next morning, free from its unexpected grounding, the ship steamed toward the botanists and ornithologists left at Kukak Bay, hoping they would find the party intact. The group of scientists they saw in the distance immediately put their worries to rest, for the naturalists were waiting patiently on the shore with piles of specimens neatly gathered at their camp. Across the water, however, at Uyak on Kodiak Island, the bear hunters were not so pleased with their days in camp. Though the five men had spotted numerous signs of bears, the luckless party had not managed actually to sight a single one. Instead, hordes of mosquitoes had descended upon them at camp, unmercifully attacking their eyes, noses, and ears so thickly that it was difficult to wave them away. Grinnell welcomed the sight of the *Elder* chugging across the water toward their camp. The futile bear hunt had hardly been worth the discomfort of the torturous plague of flies, and the hunters boarded the luxurious ship, glad for relief from the drone of mosquitoes about their heads.

There were no pesky insects in the peaceful waters along the Alaska Peninsula as the *Elder* puffed westward toward the Shumagin Islands, the first stop on the long trip to Siberia. Rumors that the expedition would eventually make its way to the Russian coast had circulated among the expeditioners for several days. Gossipers hinted that it was Mrs. Harriman who wanted to see the strange land of the Russians for herself and had persuaded her adventurous husband to make the dangerous trek across the Bering Sea. It surely would not have taken a strong argument, however, to convince the railroad magnate to venture to the Siberian coast. A bridge or tunnel could allow railway passage across the strait to Siberia, making possible a round-the-world railroad route. In a year or two, Harriman must have known, the Trans-Siberian Railroad would be complete and a connection with the Russian line could radically transform the trade between the two continents. Indeed, the entire face of Alaska could be altered if the route became a reality. Having won his bear, Harriman could once again turn his vision to the railroad world, and the engineering possibilities signaled by the White Pass Railroad probably remained vividly in his mind as the ship churned through the water toward the rocky islands of the Bering Sea.

At the time, Jules Verne's science fiction fantasies filled library shelves across the country. In an era of grand engineering ideas, booming industrialization, and astounding new inventions, a Bering Strait connection did not seem preposterous. Only a few years later, Harriman would travel to Japan, ostensibly on a diplomatic mission, with the ulterior motive of assessing the country's railroads for a round-the-world line. Railroads were rapidly making the globe smaller, in a sense, as remote areas became accessible to large business interests. A rail line could revolutionize the entire region of Alaska in a matter of months, bringing in new commerce, and unlocking "Seward's Icebox" for the voracious phase capitalist even more than the whaler and the steamer had done.

CHAPTER TEN

THE SIBERIAN CONNECTION

Eerie mirages altered the shoreline as Harriman's elite coterie continued their voyage northward during early July. The spectators on the *Elder* saw islands hovering in the sky overhead as if the world had become inverted. Like some collective hallucination, strange forms took shape and then vanished on the shoreline. While his companions fancied they saw medieval castles on the mirage-altered shore, Brewer tried desperately to reconstruct his Fourth of July speech from memory: Mrs. Harriman had been deeply moved by his oration and wanted the lecture preserved in print. Surprised at all the fuss over his talk, the unpretentious Brewer spent the day piecing together his original ideas.

As darkness closed around the *Elder*, the pilot approached the Shumagin Islands cautiously. In the obscurity of night the rocky islands spelled danger, but the ship eased its way around craggy Popof Island to reach anchorage at Sand Point, where Ritter, Saunders, Palache, Kincaid, and Kelly decided to remain behind with a launch for ten days. The scientists would camp at an old village, constructed years earlier when the government had established a station to restrict sealing. Now only one man inhabited the lonely locale of empty houses and hotels. After dropping the five scientists at the isolated point, Harriman prepared to push on toward the Bering Sea the next morning, but two gold miners, drunk on the cheap whiskey notorious in the Alaskan territory, made the departure less than tranquil. Yelling to the expeditioners, the two rowdy men tried to board the *Elder*, and when the crew prevented them, they followed the Harriman ship in a tiny rowboat, shouting angrily. Before long, however, their mood changed, and the sounds of drunken singing trailed across the water as the ship made its way among coastal islands. A cloudless day revealed the two grand peaks of Pavlof Volcano to the sightseers on deck, and all day magnificent views appeared along the shoreline. Merriam "felt more deeply than ever how little the magnificent mountains of the Alaska Peninsula and Aleutian Islands are known and how few people really have any true idea of their grandeur."

John Burroughs interrupted Merriam's reverie with thoughts other than those of the sublime mountains on which Merriam raptly gazed. Burroughs sheepishly showed Merriam a sparrow's nest and four eggs that he had taken at Sand Point and confiscated for his own, despite his previous loud complaints about the ornithologists' bird hunting. The old naturalist wanted to make a secret deal with Merriam. He proposed that the scientist give him, in exchange for the nest and eggs, the skin of a golden-crowned sparrow that Merriam had shot earlier and Burroughs had quietly envied. To the naturalist's great embarrassment, an amused Merriam boisterously announced the proposal to a crowd of scientists, who teased Burroughs unmercifully for his duplicity.

Fog rolled across the water late in the afternoon, but the tall summit of Mount Shishaldin still rose high above the gray mist. An entire series of towering mountain peaks became visible, and Muir found "grim, unmanageable strength in every feature." The ship anchored in the fog on Unalaska Island, where, at Dutch Harbor, a new town had sprung up to serve the needs of the North American Commercial Company, which dealt in the island's fur seals. The *Elder* took on water and coal for the long voyage northward through the treacherous passages of the Bering sea.

Merriam, who seemed to have old acquaintances scattered all across Alaska, took advantage of the brief stop to visit yet another friend of his in the region. Taking Gilbert and Dall with him, Merriam called on J. Stanley-Brown, the head of the North American Commercial Company at Dutch Harbor. Brown's business, along with the constant stream of transient gold miners, had made Unalaska Island a busy site. From the decks of steamships poured hordes of tired, disillusioned prospectors fresh from the Yukon territory. But, like gambling addicts in an icy casino, most of the men continued to try their luck. The Nome area, some hoped, would yield the pay-dirt that had remained elusive in the Klondike. Others—the lucky ones—headed back to the States with their pockets filled with gold dust. Rumors held that a rusty old ship anchored at Dutch Harbor bore as much as four hundred thousand dollars' worth of gold dust! Gossip surrounding the return of the miners quickly filled the town, and violence and theft in the anarchic gold region made the luckier miners fear for the safety of their new wealth. Dellenbaugh heard that "one man sent to San Francisco for two armed men to guard his gold every minute and several [miners] had a stateroom together with their gold in it, each guarding in turn." The prospectors fascinated Dellenbaugh, who had himself known the lure of adventure in the wilder days of the western frontier. The artist talked freely with the miners, most of whom, he learned, had wearied of their costly dreams. Only five of every hundred miners, he was told, had earned anything at all in their long months of prospecting, but yet another boatload of miners was expected to dock at Unalaska the next morning, and its reputed cargo of "several millions" in gold was the bait that lured the tired miners endlessly.

Tales of the gold miners' struggles sent shivers down John Burroughs' spine. As far as he was concerned, even the relatively mild upcoming voyage to Siberia seemed dreadful—the placid Unalaska Island was wilderness enough for the old naturalist. In secrecy, he packed a small satchel and strolled non-chalantly down the gangplank after quietly "checking out" on the ship's pegboard. During Burroughs' leisurely morning walks on the island, he had discovered a pleasant woman, who had won his heart by offering him fresh eggs for breakfast. He had sought fresh eggs ever since the visit at Sitka, and now he planned to "board" at his new friend's simple home while the expedition continued on its way to Siberia. But Muir and Keeler, who had been strolling about town, apprehended their friend as he descended the gangplank. Muir immediately challenged Burroughs: "Where are you going with that grip, Johnny?" A furtive Burroughs shuffled his feet and tried to avoid a direct answer. He finally admitted his plans, explaining that he "didn't like to go into those tempestuous waters." Muir, however, was relentless. Why, the Bering Sea, he chortled, was "like a millpond. The best part of our trip is up there—seeing the fur seal on their breeding grounds on the Pribilof Islands, seeing the Eskimo at Port Clarence! Come along," he insisted. "You can't miss it!" While Keeler grabbed Burroughs' suitcase, Muir accompanied his bewildered friend back to his stateroom, assuring him all the way that this was a trip that could not be missed. Ridgway and Starks had decided to remain behind at Unalaska to collect birds and flowers. The demands of their profession gave the scientists a more legitimate reason to stay, thought Muir, than the feeble excuses of Burroughs, who had become the irascible Scot's favorite whipping boy.

So it was that when the *Elder* pulled away from the Unalaska, which Merriam had described as "a veritable flower garden," Burroughs stood sadly on the deck staring wistfully back at the little village where he longed to spend a respite from the churning seas of the Bering Strait. But the ship plunged on toward the foreboding volcanic Bogoslof Islands, where several geologists and botanists wanted to lower a launch and venture into the breeding grounds of sea lions. Led by Harriman, the launch carried eight men to the rocky, desolate island, where the massive sea lions endured long months of

Bogoslof (Bering Sea), by E. S. Curtis. Inset: Bogoslof, by Captain
W. D. Doran.

battling through their mating rituals. Amid the "bellowing and
roaring" of the animals, the men waded to the island, pulling
the launch to shore behind them. The large bulls, sensing a
rude intrusion, charged down the rocks to challenge their visitors, but the party bravely stood their ground, their cameras
aimed at the bulky beasts. Merriam impulsively decided to
wage his own attack, and the salty old biologist suddenly ran
at the bulls full speed. The animals, terrified at the sight of the
two-legged creature hurtling toward them, lunged into the sea.

Fisher had joined the party not to investigate the seals but
to bring back specimens of the thousands of breeding murres
that clustered on the steep cliffs above the shoreline. Shooting
the birds was hardly difficult; the ornithologist had only to fire
his gun into the cliffs to secure his prey. Then great clouds of
murres rose into the air in spectacular flights, leaving behind
many birds that fell dead to the shore. Green eggs crashed
down behind them from their nests and smashed against the
rocks. The thunder of thousands of bird wings muted even

the loud bellowing of the sea lions, who were rushing about the rocks in a panic over the blasts of Fisher's rifle. When the killing was over, Fisher scooped up thirty birds to add to the collection he had stored in boxes on the *Elder*, and the mammalogists brought two baby sea lions back in the boat. To Muir's thinking, the "handsome little fellows were stolen from their mothers," and he winced at the rapacious disruption his colleagues had caused on the little island. After the party returned, Merriam gave a lecture on the seal rookeries, since the next day the ship would arrive at the Pribilof Islands to visit a government-controlled rookery. There, seals were "harvested" by the North American Commercial Company.

Waking in the morning to rough seas, Dellenbaugh remembered the little vial of coca extract he had acquired back at New York's Century Club. Mixed with a little whiskey, he thought, perhaps the medicine would settle his queasy stomach. Within two minutes, however, the strange combination had the opposite effect and Dellenbaugh hurriedly sought the ship's rail to "pay his respects to the broad waters." Below deck, Burroughs fared even worse. The old naturalist groaned in his bed as the ship pitched and rolled. A guilt-ridden Charles Keeler read Wordsworth to his friend to take his mind off the vertigo caused by the rough seas. Down the hall, Gifford had stoically locked himself in his stateroom to suffer through his seasickness alone.

After noon, however, the seas settled enough so that most of Harriman's passengers felt up to the short paddle to St. Paul Island to visit the fur seal rookeries. It was a rare privilege to view the government-regulated rookeries, and Harriman had secured a special permit from the Secretary of the Treasury to allow a tour by the expedition. Two men from the North American Commercial Company led the expeditioners over the island's flower-covered hills to reach the rookeries, where the seals bred, and the company killed the quota allowed them by the government.

Merriam caught his breath in dismay at the sight of the fur seals scattered on the rocks. Since his last visit to the area in

1891, when he had served on an international commission to regulate sealing, the population of the animals had shrunk drastically—less than a quarter of their previous number remained! Nervous bulls, having battled for months over the female "harems," Merriam wrote later, "roared savagely, blew steam from their nostrils and rushed at us when we ventured too near. . . ." The young males, who provided the most commercially desirable skins, had been slaughtered by the company for years and their skinless carcasses were left to decay. Such reckless policies had taken a devastating toll of the animals. Since the number of seals had diminished so rapidly, government regulations, for the most part written by Merriam himself, had stipulated that only fifteen thousand seals a year could be sacrificed, compared to the one hundred thousand the companies had slaughtered several years before. Jordan, the *Elder's* pilot, politely suggested that the government brand the female seals "so as to render their skins valueless" and ensure the survival of the species. The rookery was a dismal sight to George Bird Grinnell, who worried that the entire region of Alaska had been sadly neglected by the government, its animals pursued greedily for commercial profit with no concern for future resources. But, ever cordial to their hosts, the expeditioners joined the company representatives at the small village that served the business. Over whiskey and cigars, they discussed the urgent need to preserve the endangered species.

Fog settled in as the Harriman guests made their way back to the ship to begin the long journey toward Siberia's Plover Bay. On board, the passengers were eating dinner when a series of rough jolts shattered the meal. Thoughts of shipwreck flashed across the faces in the dining hall. Leaving half-eaten dinners behind, the alarmed scientists rushed to the deck, heedless of the steward's pleas that they remain seated. Merriam, certain that the boat's bottom had been smashed by the heavy impact, sought out the engineer for reassurance that the ship could not possibly sink. The crowd congregated on deck, grasping at the rail as the *Elder* gave two more violent bumps before the pilot freed it from the rock on which it had lodged in the heavy fog. Soon the ship, miraculously undamaged by the collision, sailed on. After the danger had passed, Merriam wrote sardonically in his diary: "A great day—the Pribilof Fur Seal rookeries and tundra gardens and the promise of a shipwreck all in one day!" But Burroughs, ever cautious and conservative, hoped that the near-accident would convince Harriman to turn around. Harriman, however, remained undaunted by the possibility of danger. Within minutes, Burroughs noticed with dismay, "he was romping with his children again as if nothing had happened." Captain Doran, however, made sure the ship announced its passage through the dangerous fog with its shrill whistle, interrupting the sleep of the passengers throughout the night.

Burroughs regretted Harriman's decision to continue to Siberia even more the following day, when rough seas kept him in bed until nightfall. Keeler, too, spent the day rather unhappily; he had learned that the expedition members were to "pay for this voyage with a vengeance." Harriman, it seemed, had requested that all photographs be given to him at the end of the trip. This way, the railroad tycoon could select those he wanted to include in the leather-bound "souvenir album" of photographs he would distribute to each of his guests. Moreover, Harriman had decided to prohibit any writing on the expedition other than approved articles. Too much advance publicity, thought Harriman, would detract from the worth of the two volumes he planned to publish. To complete the poet's doleful day, gray fog and rain closed around the ship as it made its way toward the Siberian coast.

The expedition reached Plover Bay on the Siberian coast the next afternoon, under the authoritative guidance of Dall, who had visited the area years before. Most of the expeditioners became excited at the sight of several boatloads of Eskimos approaching the *Elder*, and cameras immediately came popping from staterooms that had emitted only seasick groans the day before. The picturesque sight of numerous Eskimos selling their wares in the waters beneath the Elder appealed to the amateur photographers among the expedition members. But as the ship, true to its original nickname, the "George W. Roller," rolled to each swell of the sea, its passengers became anxious to reach the sordid-looking village at Plover Bay. Merriam impatiently set out in a launch. On landing, his first reaction was that the Siberian women "were not attractive," and the male Eskimos, he found, bore horrible sores on the tops of their bald heads. Syphilis scars badly marked the natives, the cruel legacy of Russian whaling expeditions and explorers who had sexually exploited them, leaving them with foreign diseases. The impact of the "civilizing influences" so touted by some expeditioners had left undeniable stains on the people of the Siberian coast. "The contact" wrote a thoughtful Muir, "with civilization of the whaler seamen sort had, of course, spoiled them."

The sight of the small settlement on Plover Bay must have been somewhat of a disappointment after the long rough hours spent to reach Siberia. About twenty-five natives, dressed in reindeer-skin parkas and sealskin boots, inhabited what Merriam called "the most barren and desolate place of its size [he] ever saw." Icy winds ripped across the rocky coast and the visitors shivered in their cloth coats. Harriman had bought two reindeer coats at Unalaska, and he stood erect in a fur coat, the envy of his companions. Bracing themselves against the strong wind, the expeditioners made their way along the beach, staring curiously at the Eskimo villagers and peering into their small summer huts of skins and canvas; in the winter, the natives constructed housing of sod to guard against harsher weather. Inside the tentlike structures, the white visitors squinted past smoky fires to discern remnants of seal and walrus, obtained on a recent hunt. Animal remains lay strewn about the strong-smelling huts. Curtis and Inverarity busily took photographs; here, they found, the adult Eskimos willingly allowed cameras to be aimed in their direction, but the children became frightened at the appearance of the unfamiliar visitors. Even the villagers' scrawny dogs bounded away in terror to the rocky hills, where, Burroughs wrote, "they sat

Ashore at Plover Bay.

Siberians at Plover Bay.

Siberians and Harrimans at Plover
Bay.

Siberians at Plover Bay.

Meeting at Plover Bay.

Eskimos in Umiaks—Port Clarence, Alaska, by E. H. Harriman.

down and howled their mournful protest." After Harriman had befriended the Eskimos with gifts of tobacco and glass beads, the scientists secured from the natives items ranging from sealskin boots to walrus-harpoon heads and ivory hooks. Devereux happily traded a knife for a pair of sheep's horns, since the natives preferred tools, tobacco, or sugar to the strange round pieces of metal the white men offered.

The novelty of Siberia wore off quickly, and the bone-chilling winds soon drove the party back to the ship. On the way to Port Clarence, directly across the Bering Strait, where the *Elder* would take on water, the sun never sank beneath the horizon, and several expeditioners sat on deck all night to relish the "midnight sun." William Dall entertained his companions with stories of the Siberian natives. The group, he said, usually killed their old women when they grew useless, and sometimes the women actually requested to be put to death. Imagine, Dall continued, what it would be like to be stranded among such people. Such an incident had actually occurred years ago, when the lone survivor of a shipwreck had struggled to shore, only to be made a slave by the Siberian natives and sold to tribes in the interior. After a long period of slavery there, the unfortunate sailor was finally sold back to the coastal tribes, where he managed to set afloat a piece of driftwood with his name carved into it, as well as the point along shore where he imagined himself to be. According to Dall's story, a ship returned to rescue the man months later from his slavery among the Siberian Chuchchis. Such stories of adventure abounded in the rough region, and Dall held his listeners spellbound by fearsome tales during the long night.

It was a relief for the expeditioners to reach the "American side" of the Bering Sea after hours of sailing. The *Elder* drew up to a port filled with ten whaling ships preparing to begin long voyages into the Arctic in August, for Port Clarence provided a way station for both outgoing whalers and the returning ships that needed to replenish their supplies. As if keeping a snobbish distance from the whaling ships, the *Elder* dropped anchor several miles from the port. Despite the distance, the scientists soon saw umiaks bobbing across the water toward them. The natives paddled out to greet the *Elder* as they had done at Plover Bay, eager to find buyers for their furs. Their faces, thought Dellenbaugh, "looked much cleaner" than the Siberians; these natives "were fairly clean for people of this class, but of course," he added, "there is a great odor of fish

Whaling Fleet—Port Clarence, by E. S. Curtis.

and seal oil about them." To Muir, the lively bands of Eskimos in their umiaks of walrus skin seemed like "a merry gypsy crowd." The natives had brought along numerous children, tiny babies, and even dogs in the vessels laden with their wares. The stretched, waterproof hides of the umiaks shone a brilliant yellow beneath the sun as the crafts skimmed easily across the water.

Though several whaling men boarded the *Elder*, Harriman would permit none of the Eskimos on board. The whaling boats' captains climbed on deck to visit the great ship, congregating in Captain Doran's cabin for strong drink and cigars, and Muir joined the group to discuss the whaling industry.

With the Whalers at Port Clarence (Harriman on right with white feather in hat).

"Eskimos on a Whaler," Port Clarence.

One captain informed the naturalist that whales abounded around the MacKenzie River's mouth; twenty-five or so whales in a season, he reported, provided a good hunting catch. Their fleet belonged to the powerful Pacific Steam Whaling Company based at Orca, which Captain Humphrey headed. Among the guests in Doran's cabin was a gold miner who was temporarily in charge of the United States' reindeer station farther north, where a reindeer trade operated between Siberia and Alaska. Far more interested in gold than reindeer, the man told the company fantastic stories of new gold discoveries at Cape Nome to the north. Along with thirteen partners, he confided he had staked out excellent diggings and taken out twenty-eight claims. In a matter of only weeks, Nome would become the focus of a new gold stampede. Harriman's guest must have been one of the first to stumble across the profitable new fields. Miners from the Yukon and the Klondike would soon desert their old claims and pour into Nome, vastly increasing its population in August of 1899.

After giving his isolated guests old newspapers to read from the United States and in turn collecting their letters to mail, Harriman lowered a launch to inspect the whaling ships for himself. Along with Devereux, Elliot, and Chaplain Nelson, the tycoon toured the rugged old ships, chatting with the salty whalers who pitted their skills against the mightiest of the ocean's creatures. Several small boats followed Harriman from the *Elder*, one bearing the Harriman family, who braved the rough surf to explore the shore. Landing at Port Clarence proved to be even more difficult than the strenuous rowing, and the women and children had to be carried ashore. It amused Muir "to see the demolishment of dignity and neat propriety." Once ashore, the passengers found Port Clarence itself easier to negotiate, though its marshy brown tundra and wet hills made it difficult to explore far from the coast. Muir, typically, struck out across the terrain and, as agile as the mountain goats he admired, explored the hills to observe their vegetation.

Gifford and Dellenbaugh chose to bargain with the natives, and they came away from their spirited negotiations clutching a pipe and a walrushide line with a harpoon head attached. Merriam's bartering was not so successful. After arguing endlessly with a cantankerous old Eskimo over a pipe, he finally realized the man was roaring drunk. The scientist backed off in disgust while the drunken native yelled at his prospective buyer to return. Kept amply supplied with cheap whiskey and diluted Chinese alcohol by the whaling men, the natives were reduced to trading their valuable furs and skins for intoxicants. A gold miner passed along a warning to one of Harriman's guests: by evening, he predicted, all of the natives would be riotously drunk. Grinnell listened angrily, determined anew to devote his life to the preservation of Indian cultures. Always ready to befriend the natives, Grinnell graciously accepted an Eskimo's offer of lunch, and bit down into a piece of rancid walrus hide. To the Eskimo's amusement, a great grimace crossed Grinnell's face at the taste of the oily skin, and Merriam, standing nearby, howled with laughter at the efforts of his colleague to chew the food.

Keeler, too, was off meeting the inhabitants of the area. There at Port Clarence, the poet discovered a miner who had not heard news of the "outside world" in two years. Information about the Spanish war and the conflict in the Philippines astonished the isolated traveler, and Keeler explained the details of the political events to his fascinated listener. While Keeler talked endlessly of the nation's news from the past two years, Curtis rambled over the decks of the whaling ships. With his keen photographer's eye, he observed the natives mingling with the seamen on board. This, he thought, could be a profitable place to spend extra time with his camera, and he arranged to eat dinner on board one of the whaling ships while the rest of the Harriman expeditioners returned to the *Elder*. One of the smaller boats was left behind for Curtis so he could row back to the ship later in the night.

Thinking that only his photographer remained on shore, Harriman returned to the *Elder* in his steam launch. Once on board, however, he discovered to his alarm that his wife and children had been left behind as well. Harriman, though exhausted from the long trip to the ship, immediately climbed back into the boat to rescue his stranded family. Dinner in the plush dining room would be delayed until Mr. Harriman returned.

CHAPTER ELEVEN

"I DON'T GIVE A DAMN
IF I NEVER SEE
ANY MORE SCENERY"

At Port Clarence, the Harriman Expedition reached the northernmost part of its long journey. When on July 13 the *Elder* turned its prow south to retrace its path homeward, most of the expedition members looked forward to arriving in Seattle in two weeks. Charles Keeler had spent the journey pining for his family in San Francisco, while both Palache and Burroughs were eager to get back to their East Coast homes. But the irascible Albert Fisher was discontented with the limited contents of the crates and boxes of stuffed birds he had collected; he had envisioned taking back many more specimens to pore over in the laboratories of the Smithsonian. In the late afternoon, when the ship docked at St. Lawrence Island for a brief stop, Fisher toted his rifle once again in a launch to hunt down still more birds. Even the rain and dense gray mist did not stop him from paddling out to the island for work with a group of his colleagues. The Harriman daughters accompanied the scientists to explore the terrain, where polar bears were reputed to roam. It was, possibly, the chance of bagging a different sort of bear that also led Harriman out to the little island, but Dellenbaugh watched the parties rowing ashore in the afternoon and wondered what the fuss was about. The barren island held "absolutely nothing to attract me," he thought, and turned his attention to his sketches.

Perhaps it was the illusions created by the mirage-ridden weeks of the long journey or the dreary day of fog and mist that made the Harriman daughters simultaneously think of polar bears when they saw two white creatures off in the distance. At the girls' excited cry of "Bears!," Merriam immediately took off by himself across the barren tundra, hauling with him only his 20-gauge shotgun. Only a "close quarter" shot, he thought, could kill one of the strange-looking beasts. For two miles he trailed the animals, catching brief glimpses of their white coats as they eluded him. But at last, one of the creatures stopped while Merriam skulked nearer and nearer, raising his gun carefully to take aim at the animal's back. With a sudden croaking, the alarmed animal saw its hunter and, as it took off toward its mate in fear, Merriam's heart sank. They were swans! A small brood of their young waddled quickly behind the two, and Merriam lowered his gun sheepishly after his wild-goose chase. There would be no stuffed polar bears to adorn the halls outside his Smithsonian office. By the time the disappointed scientist rejoined his party, Dr. Morris was showing off the young swans he had killed on the opposite side of the island. Five of the soft downy creatures lay sprawled on the shore, ready for the taxidermists' knives.

Merriam's penchant for a good story did not stop even when the story was at his own expense, and his companions back on board the ship laughed heartily along with him at the outcome of his bear-hunting trip. Fernow used the incident to pen a verse about Merriam in a poem about the expeditioners:

> Of all the merriest, Merriam shines!
> Geese, swans, or bears? It's fine he opines
> Whatever it is! His rubicund face,
> Sees beauty in seals, in sea-lions' grace.

Later that evening, Fuertes delivered an after-dinner lecture, complete with musical imitations, on the songs of Alaska's birds, which had so intrigued the young artist throughout the voyage. Although his colleague Fisher was not completely happy with the findings of the trip, Fuertes felt satisfied with his own work on the expedition. He counted ninety-five bird skins in his collection and, as he had followed the advice of his mentor Elliott Coues, sketches and paintings filled his portfolio. He was especially pleased with the many friends he had made. "I've had good luck," he wrote home, "and a bully time."

Fuertes had left St. Lawrence Island happy with the selection of birds they had acquired on the site, but the expedition's next stop, at Hall Island, would prove to be even more rewarding for the ornithologists. Fisher, who by late June had felt "perfectly ready to go back when the time comes," ambitiously set out his traps on St. Matthew Island the next day, with Curtis helping him. While the two men set the traps on shore, Brewer meditated on the ship's deck: "There is something wierd [*sic*] in this," he scrawled, "as we are lying at anchor, by these desolate . . . islands, so near the middle of the out-of-the-way but vast Bering Sea, with mysteries of fogs and gloom." So remote were the islands, he noticed, that even wild birds flew fearlessly about the ship, unaccustomed to the sort of danger posed by ornithologists with guns. Perhaps that was why Fuertes arrived back at the *Elder* late at night with his small boat loaded with dead birds from nearby Hall Island. As he boarded the ship Fuertes was still amazed at the evening's experience. Rare species of sea birds had covered the cliffs of the island. It was, he wrote, "truly the most wonderful sight I've ever seen. Thousands and thousands of birds— tame to stupidity, seated on every little ledge or projection— from the size of sandpipers up to a great white gull that spreads five feet—all the time coming and going, screaming, croaking, peeping . . ." From the top of the cliff Fuertes had caught the birds in his bare hands, one after the other. The long daylit evening had been an ornithologist's dream.

But while the region provided an ecstatic dream for Fuertes, there was something sinister and nightmarish about the remote islands to Brewer, and a story told by Dall that night confirmed Brewer's eerie feelings. The ruin of an old sod house on St. Matthew Island, it seemed, had been the site of a ghastly murder many years ago when a hunting party of four Russians had camped throughout a long winter. By the spring only one hunter remained, and he told a suspicious story to his rescuers of how his companions had drifted out into the Bering Sea on ice floes and never returned. But legend had it, Dall continued to a hushed audience, that actually the harsh winter in the midst of the Bering Sea had driven the lone survivor mad, and he had killed and eaten his companions. In the thick, gloomy fog, Dall's audience listened to the rough waves lap against the ship, and found the story easy to believe.

Though Fuertes' imagination often had a free rein during his hikes in the Alaskan woods, the excitement of the bird finds were uppermost on his mind. After sleeping only briefly, at four the next morning he rowed with Fisher toward St. Matthew Island to check the traps his companion had set out a few hours earlier. Sitting in the bow of the small rowboat, Fuertes thought he glimpsed a Bonaparte Gull through the mist. Fisher, busily rowing, barely glanced up to bark an order. "Pot that," he said. "It's a Sabine." Fuertes raised his rifle to shoot the rare bird and indeed, his older colleague's guess proved correct when they examined the dead creature. With the graceful Sabine specimen, Fuertes thought approvingly, "we had the finest bird the trip had yet produced." With its grayish blue head, white neck, and rosy breast, the dead gull would make a lovely contribution to Fuertes' portfolio.

Later a larger party with exploring and bear hunting in mind joined the two ornithologists on St. Matthew Island. Merriam's polar bear "sighting" had, it seemed, triggered the imaginations of the many frustrated hunters. Perhaps on the mid-sea islands, where polar bears supposedly were often stranded by melting ice floes, they could bag a hunter's trophy. The snow-white skin of a giant polar bear could easily

grace either the halls of Washington's scientific establishments
or the private den of a proud hunter. Even Dellenbaugh, after
he had mounted photographs for Mrs. Harriman's album, toted
his rifle as well as his sketching equipment to shore. But the
only game were the island's numerous birds, and Fuertes,
Fisher, and Trudeau could not seem to gather enough specimens. The sounds of their popping guns followed Dellenbaugh and Keeler as they wandered over to explore the
infamous ruins of the reputed cannibal's sod house, but all
they found at the grisly site were rotten timbers extending
back into a bluff.

At noon, the *Elder*'s whistle signaled an end to the visit at
St. Matthew Island, and Muir sighed when he saw the ornithologists proudly hauling their birds on deck. In addition to
a beautiful white gull, he counted a "snowy owl, and two
young blue foxes and one old one, the mother; the pitiful things
were laid out on the wet deck." His companions, he noticed,
had also stolen bird eggs from their nests and killed the par-
ents while "many little birds were left to starve." Alaska's
dwindling resources and wildlife, he thought, could not afford
many more such attacks in the name of science. Muir's own
personal studies of the region had never interrupted the har-
monious cycles of nature or involved violence against his fellow wilderness creatures, whom he regarded as having equal
rights to life. Though the expedition had allowed him to visit
Alaskan lands he had never before seen, he would be glad to
return to his solitary style of exploration.

Under murky, overcast skies and through rough seas that
left a miserable Burroughs confined to his bed, the *Elder* cut a
long path down the Bering Sea back to Unalaska while its
passengers entertained themselves spotting whales from the
deck or admiring each other's photographs, which Curtis developed daily. Poetry and humorous verses often adorned the
bulletin board of the lecture hall, providing reading matter that
commented on the writers' companions or on a day's experience, and passersby stopped to amuse themselves at the cre-
ative attempts of their fellow passengers.

The sight of Dutch Harbor on the Alaska Peninsula was a welcome one the next morning. The fog had given way to a balmy breeze that cleared the air of the peaceful bay at Unalaska, and sunlight streaked across the green hills surrounding the harbor. John Burroughs must have feasted his eyes on the idyllic scene. Glad to see land again, Dellenbaugh pulled out his paintbrushes to put the time ashore to good use. He had not been painting long before a lonely gold miner interrupted his work and, as usual, the artist politely chatted with his visitor. For two years, Dellenbaugh learned, the miner had nourished his dreams of finding gold in the Yukon, but the rough area and the hardships along the path to wealth had left too many scars. With just enough money for a return fare, the prospector was heading back to join his family in New York. A ship in the Unalaska harbor, Dellenbaugh had noticed, seemed actually to groan with its load as disenchanted gold miners jammed every cranny for the voyage to San Francisco. The artist's new acquaintance was searching the beach for a last-minute souvenir to bring his family, and Dellenbaugh reached into his pocket to draw out several jasper pebbles he had collected on a Bering Sea island. "Your family will be glad to see you," he said, smiling, as he gave the miner the rocks. "Yes, and I'll be glad to see my family," said the weary prospector. Feeling very virtuous, Dellenbaugh watched the man stroll down the beach, and hoped that the "gold fever" would not strike him again. After only a couple of hours in port, fog closed in again around the ship, and it slowly departed through the gray haze. A loud whistle announced its presence until a solid curtain of fog forced Doran to drop anchor in the night.

After anchorage, it was Keeler's turn to address the expedition in the lecture hall, and he chose once again to confront his companions with controversy. Keeler's Fourth of July talk on United States intervention in the Philippines may still have been in the minds of his audience when the young poet took the podium, but this time, it was a variation of Darwinism he expounded, using the coloration of Alaskan birds to illustrate his point. Like a "pyramid standing on a good firm base," he explained, heredity ensured slow, conservative change, and most variations in animal life would fade. "Only a few" that happened to suit their environment best would survive, Keeler concluded.

Keeler's scientific message became quickly submerged in the committee meetings that immediately followed, however, and John Muir grumbled in typical fashion at the plans for the two volumes to be written on the expedition. The long conference, he thought, involved "much twaddle about a grand scientific monument of this trip. . . . Much ado about little," he scoffed. "Game-hunting, the chief aim, has been unsuccessful. The rest of the story will be mere reconnaissance." But most of his company did not agree, and when Harriman followed Keeler to the podium he told his grateful audience of scientists and artists, "Let nobody think anything about repaying Mr. Harriman. Mr. Harriman," he assured his listeners, referring to himself in the third person, "has been amply repaid by the pleasure he has had out of it." If, as Keeler had just implied, only the fittest survived, then perhaps Harriman felt it the responsibility of corporate survivors like himself to act as generous benefactors for those still struggling.

But Harriman's pleasure in the expedition was beginning to wane. The railroad tycoon anxiously anticipated a return to his beloved world of business intrigues. When the *Elder* passed the majestic Fairweather Range during the last days of the expedition, Harriman sat with his wife on the opposite side of the ship's deck. Dazzled by the magnificent display of the mountain range, Merriam hurried to urge his host to join the spectators. Seeing the Harrimans lounging in deck chairs that faced seaward, he beckoned to them and called out, "You are missing the most glorious scenery of the whole trip!" But Harriman was not to be budged. Sullen from the long weeks without communication with his business colleagues, he snarled at Merriam, "I don't give a damn if I never see any more scenery!"

Pleasure had yielded to frustration and boredom for the strong-willed Harriman, who thrived on the brisk competition and power of the nation's railroad builders. The scientists' cautious methods frustrated him, and he later told a reporter that "the scientists have a way of reducing everything to an exact point and would probably manage a railroad from some such exact basis, only it would take them a century to go from Chicago to St. Louis." Weeks of leisurely cruising in Alaskan waters and catering to the scientists had whetted his appetite for the energetic pace of Wall Street.

He would not have long to wait. The days passed quickly for the remainder of the expedition. There were few stops scheduled as the *Elder*, pushed on by the impatience of its passengers, steamed toward Seattle. Brewer whittled away the hours on deck by composing riddles to post on the bulletin board. His feelings for the surreality of the voyage's latter days emerged in one strange piece of verse he tacked to the poetry board:

> A headless man had a letter to write
> Twas read by one who lost his sight,
> The drunk repeated it word for word,
> And he was deaf who listened and heard.

Weeks on the Arctic seas had stirred the sensitive old botanist's unconscious memories and troublesome dreams disturbed his sleep. One night, Brewer's mother, who had been dead for thirty-four years, appeared in one of his dreams with a new husband, about whom she had not told her family. Amazed at the revelation, Brewer woke with a start, and the strange dream haunted him throughout the day. An otherworldliness seemed to hover about the ship as the expedition sailed in its final days, and Brewer found, a man's clear vision was prevented by a mist that was "so fine no drops can be seen. . . . [It] adheres to my spectacles," he wrote, "as a film."

At Sand Point, the *Elder* picked up the party of scientists headed by Palache who had chosen to remain at the isolated island for botanizing. Palache himself was sprawled out lazily in the grass beside a shallow stream, Tom Sawyer–style, and a string of trout lay beside him in the shade of a willow tree.

Merriam carefully measured the trout his colleague had caught—three or four feet each, he recorded. The campers had enjoyed "all the trout they could eat," but their stay in camp at Sand Point had not been without excitement. Earthquakes had rocked them violently one night, and the scientists had wakened to an awesome roar as the deserted hotel where they "camped" for ten days shook fearsomely. But the danger had been worth it, they thought, and that evening Ritter, Saunders, and Kincaid reported on the insects and marine life they had collected during their stay.

Three thousand insects would accompany the entomologist Kincaid back to the University of Washington, and he excitedly discussed his tiny finds with an audience that sensed his professional commitment to his work and shared his enthusiasm. Kincaid's talk on his insects, thought Muir, "was one of the very best of the trip. He has genius," added the admiring naturalist, "and will be heard of later. . . ." His mimicry of insects and dramatizations of the lives of bees and butterflies dazzled the audience and, for his astonishing performance, which rightly belonged in Shakespeare's *A Midsummer Night's Dream*, "all cheered him heartily."

The *Elder* steamed toward Kodiak Island to replenish its supply of coal and water, while its passengers congregated to watch the Pavlof Volcano smoking in the distance. Brewer, ever conscious of the climate, informed his company that the sun's aspect had changed greatly since they had been at the site eleven days ago, a change that was particularly visible in the sun's rays on the volcanic peak.

Back at Kodiak Island, Harriman wanted to touch shore again on the island that had given him his prized bear. The *Elder* anchored and, despite the rain and fog, Harriman and several scientists steamed ashore in a launch. Perhaps sensing a last opportunity to collect specimens, the experts scattered with their rifles and traps. The young Harriman girls, however, saw a final chance to explore the charming island that had so beguiled the entire group earlier. With Merriam, they climbed the mountain rising gently behind the town for a last view of the scene. While his female company collected plants and flowers, Merriam shot a magpie, a sparrow, and a warbler.

With the various missions completed, Cornelia Harriman looked forward especially to the gathering at the Kodiak wharf that had been scheduled for ten-thirty that morning. It was her fifteenth birthday and, after a group photograph by Curtis, a celebration of the day would take place at lunch. She hurried down the hill along with her relatives and friends to the pier, but the schedule of the expedition, she had already learned, seemed to differ from that of the clock. The entire group finally congregated before Curtis' camera at noon. Afterward the party found a table laden with wild roses and rich cakes blazing with fifteen candles. The baker, according to instructions, had hidden coins in some of the cakes and Cornelia's youngest brother, Roland, bit into the piece that yielded the biggest prize—a two-and-a-half-dollar gold coin. He was learning early that pleasure could be combined with profit.

After the gala lunch, launches set out for a cruise to a nearby island east of Kodiak, where a fox farm and a salmon cannery operated. Guided by M. L. Washburn of the North American Commercial Company, the expeditioners toured the businesses run by Washburn's company. Merriam, however, left no doubts in Washburn's mind about the risks such fox farms—despite their precautions—ran of eliminating the Alaskan population of blue foxes. The company slaughtered the male foxes at a rate that considered only profit, and now, Merriam noted, "they have not enough males to fertilize the females and as a consequence many of the females have no pups." After listening to Merriam's expert knowledge on the subject, Washburn satisfied his guest that his business would alter its policies. It was hard for Dellenbaugh, however, to believe that a scarcity of foxes could occur. The animals, it seemed, were "getting so numerous on the island that they now come around the buildings." As the skins were worth about twenty dollars apiece on the market, Washburn's company made a healthy profit on its fox breeding. Feed for the animals was free—they dried the region's abundant salmon to provide a continual food supply before the fox furs were "harvested" for their contribution to the world of high fashion. Alaska's exploitable resources were seemingly infinite in their diversity.

It was evening before the expeditioners returned from the island, and Dellenbaugh sketched the view of moonlight rising on the tranquil water as the boat made its way toward Kodiak. Champagne, uncorked in honor of Cornelia Harriman's birthday, greeted the troops as they boarded the *Elder*, but George Bird Grinnell was not in a festive mood. The governmental neglect of Alaska was an abomination, he thought. The fox farm was only a small indication of the direction in which the entire region was headed if measures were not taken immediately to remedy the dire situation. "The fur seals are practically gone, the salmon are going fast. The fur trade is destroyed, no sea otter and small fur is going, the deer are disappearing," he jotted quickly in his diary. Perhaps his magazine, *Forest and Stream*, which had already launched the fast-growing Audubon Society, could take a more vociferous stance against the corporate pillage taking place in the Alaskan wilderness. As for the region's natives, their exploitation and decline was the saddest of all.

The following day, while the ship proceeded to pick up Ridgway and Kincaid at Cook Inlet, the gong sounded for the evening's lecture, and Dellenbaugh nervously sorted his notes for his talk. Little did he know that a conspiracy was afoot to disrupt his lecture. Mrs. Harriman, at the onset of the expedition, had asked Dellenbaugh to perform the duty of enlivening the often tedious evening lectures that the Committee on Entertainment sponsored. The artist had conscientiously responded to his hostess' request by asking questions after each lecture, "partly," he wrote, "for information, partly to break the dead silence that settled down when the speaker stopped." When it came his turn to lecture, however, Dellenbaugh dis-

covered a veritable "conspiracy to flood me with questions." His talk had barely begun before Gannett and Coville deluged him with ridiculous queries. "When I spoke of the wind blowing toward me," the artist wrote, "Gannett asked if wind did *always* blow toward one." And so the good-natured harassment kept up throughout his talk, to the great amusement of the audience, who were relieved that this lecture bid fair to match Fuertes' bird-warbling and Kincaid's "dance of the bumble bee."

After *all* the expeditioners were collected again the next day, the *Elder* clipped along across the coastal waters, and only John Burroughs felt the rapid uneven motion sufficiently to remain securely rooted in his chair on the deck. Meanwhile, his companions, now relentless in their search for amusement and feeling the imminent end of the grand expedition, prepared to celebrate Mrs. Harriman's birthday on July 22 in a luxurious style. Dellenbaugh busied himself making a giant candle to surprise his hostess. He presented his elaborate, handmade candle to the ship's steward early in the evening, and watched with satisfaction after dinner as a large cake bearing his flaming candle was brought to Mrs. Harriman. Champagne filled the delicate crystal glasses at each place, and the old family friend, Dr. Morris, proposed a rousing toast to the health of Mrs. Harriman. Poetry readings, music, dances, and the unveiling of new paintings completed for the occasion followed the gala dinner. J. Stanley-Brown, grateful for Harriman's generous hospitality, presented Mrs. Harriman with an unusual local gift: a cribbage board made from a walrus tusk lay beneath gay wrapping paper, neatly combining the Alaskan resources with the pastimes of the American upper class.

While its very civilized passengers danced and drank champagne, the ship proceeded once again toward Yakutat Bay to deposit Indian Jim, the colorful guide who had led the expedition safely through numerous unknown passages. After the night of celebration, only a few expeditioners woke early to a cloudy morning at the Malaspina Glacier. William Brewer, however, was awake before eight o'clock to record the temperature and the day's barometric pressure in his scientific journal. Always attuned to the mathematical and statistical ramifications of the climate that often befogged his glasses, the botanist recorded the pertinent figures in the pages of his scientific diary. Earlier, he had been fascinated by the fact that his colleague William Dall, on one of his numerous trips to Alaska, had once picked a cabbage leaf that measured three feet in length. At the close of the nineteenth century, the importance of recording in minute detail the most trivial of facts seemed crucial to establishing the credibility of science as a profession. Indeed, on the rail trip up to White Pass, Merriam had jotted down every Latin name of the animals he spotted along the way, underlining them carefully, while seemingly oblivious to the human drama of the gold miners around him.

Indian Jim, however, with his eyepatch, loose clothing, and geographical expertise, bore no diaries or papers with him on the journey northward. He now looked forward to the *Elder*'s next stop, which would return him to his family in Yakutat Bay. Beneath a sky that the meticulous Brewer described as "hazy" and cloudy, the Indian studied the horizon for signs of his native village. The clouds cleared as if in expectation of his arrival, and Mount St. Elias loomed against the morning sky, clearly visible from a distance of sixty miles. As the ship's passengers watched from the rail of the deck, Indian canoes by the dozens began racing toward the ship. In a classic understatement, Dellenbaugh speculated that the Indians "were probably anxious to see Jim again."

It was a homecoming in grand style, and as their loyal Indian guide took a boat to shore to rejoin his family and friends, Dellenbaugh maintained a careful artistic eye on the scene, and judged that the Indians of Yakutat Bay "do not compare in picturesqueness with the Eskimos." But despite the Indians' "artistic shortcomings," Dellenbaugh sketched the homecoming, while others on board the *Elder* took several boats to shore to hunt again. These gun-toters soon paddled back to the ship, and Muir noted that they "returned baffled and disgusted." Muir's sardonic tone overtook him and he penned a parody of the hunters in his journal:

> No Bears, no bears, O Lord!
> No Bears shot! What have thy servants done!

For the hunters, the last chance at shooting a bear like Harriman's apparently had passed, and it seemed that neither polar bear nor brown bear were terribly amenable to giving their lives for the honor of the Smithsonian Institution.

The expeditioners bade farewell to their Indian friends as they cruised out of Yakutat Bay for a long day of clear sailing along the coastline. The morning of July 24 promised one of the fairest days the expedition had yet seen, and the spectacle of the coastal Fairweather Range did not disappoint the passengers in the early afternoon. With its sharp, snowy peaks outlined against a crystal blue sky, the mountain range displayed its beauties for the *Elder*'s passengers. But for Dellenbaugh, who emerged on deck bearing his paintbrushes and paper, the sight was only frustrating. The sudden lurches due to the rough seas made painting impossible and he soon abandoned his efforts. One sudden jolt, in fact, caused a crewhand to tumble down a hatch on the ship, breaking several ribs when he landed. Despite the heavy waves, Gifford had managed to rise early enough to catch the sunrise over the grand Mount St. Elias, and the artist had busily sketched the ethereal vision before it escaped him and the high seas prevailed. John Muir's fellow passengers may not have appreciated the grand view as much as he and Gifford did; indeed, his breath was taken away by the clarity of the view of the Fairweather Range, which glowed, he wrote, "in its robes of snow and ice with ineffable beauty and glory of light." Only the topmost peak of Mount St. Elias lay hidden in the clouds, but below a yellow light illuminated the majestic mountain. The views that the expedition members had celebrated on the voyage northward could not compare with the unobscured magnificence in which the peak revealed itself in late July. Here, Muir thought, was the "most glorious [scene] of all the trip, surpassing even the scenery of Prince William Sound." Merriam agreed with his colleague. It was, he wrote, "a fitting conclusion to such a grand trip."

A nearly round moon shone down on the still Alaskan waters when Daniel Elliot began his scheduled lecture, drawing the passengers from the deck, where they had been practicing college cheers from their various alma maters, into the lecture hall. But the almost hysterical headiness and gaiety of the night would not subside into the usual decorum of the *Elder*'s scientific talks. Though, as Grinnell carefully noted, only water had been drunk at dinner, Elliot's audience grew riotously merry. The group interrupted his dignified talk on Somaliland with cries of "What's the matter with Elliot? He's all right! Who's all right? Elliot! What's the matter . . ." and on and on they continued, while a baffled Daniel Elliot tried to deliver his lecture with some residue of dignity. John Muir had already attributed the spontaneous merriment to an "interesting result of ice action" and the "proximity of the glaciers." But, more than likely, the effects of weeks of close living aboard ship were beginning to break through the thin veneer of propriety and self-conscious decorum, as the scientists "began to loosen the tension of nerves by shouting, joking, cutting up generally." Said Muir, "we all felt more free and friendly and reserve went to the winds," as the scientists danced jigs and sang lustily beneath the full moon on their way toward their last Alaskan stop, at Juneau.

CHAPTER TWELVE

"THE TAKING OF
THE TOTEMS"

Frederick Dellenbaugh awoke in his stateroom on the morning of July 25 with the underside of the Juneau dock outside his porthole. The tide had fallen so low that even a ship the size of the *Elder* sank far beneath the pier, and the passengers disembarked on a gangplank extended from the hurricane deck. Grinnell looked around him at the green sloping hills surrounding the town—ruined, he thought, by the ugly stumps of withered trees that bristled on the rolling mountains. Dellenbaugh immediately went out to buy a paper for the most recent news. At their last stop, at Dutch Harbor, one of the expeditioners had found a month-old newspaper and it was passed eagerly from hand to hand among the passengers. Starved for reading material, the scientists read the advertisements as carefully as they did the news articles.

His newspaper tucked under his arm, Dellenbaugh joined Gifford and Keeler for a shopping trip through Juneau's thriving business district. Prices seemed extremely reasonable, and the men carried back several Indian baskets with them. The warm, sunny day was conducive to exploring, and as they made their way along the plank sidewalks, a distinguished-

The Deserted Village—Cape Fox, by E. S. Curtis.

looking gentleman stopped them to chat. Curious about the refined man, the three expeditioners visited his home and were surprised to find it "well furnished and thoroughly civilized." Indeed, the sounds of a canary came chirping from a back room. After sailing the remote regions of the Bering Sea, they found such amenities strange in a new frontier town along the rough coast. Muir thought the town far quieter than it had been on their first visit, since the miners must be "all away getting gold or trying to get it, mostly for poor or vicious uses." Like his companion Muir, Dellenbaugh soon took off by himself to explore the hills beyond the town. A deep gorge opened before him as he climbed, revealing a mine excavated below. "Evidences of mining," he found, "were everywhere," carving deep holes into the hills that the timbering industry had already decimated. But the *Elder*'s whistle, demanding that the passengers reboard for a coal stop at Douglas Island, drew his thoughts to the ship again.

After a brief pause, the great ship sailed through coastal passages made silvery by the full moon. Harriman, remem-

bering a deserted Indian village that Dellenbaugh had learned about from one of his gold-miner acquaintances, approached the artist to ask the details of the story. The tale of the ghostly village piqued his interest; it would make a fascinating stop if it were not too far off the *Elder*'s route. Dellenbaugh pulled out the crude penciled map that the old miner had drawn for him, and he and Harriman compared the sketch with an official map of the area. An inlet called Foggy Bay appeared on both maps; nearby, they surmised, a site labeled "Cape Fox Village" on the official map must be the deserted village in question. The next morning, the ship edged along the coast below Wrangell, and presently weathered totem poles and small, strangely decorated Indian cabins appeared on a distant beach. No signs of life stirred on the desolate spot. The expeditioners lowered launches immediately and soon arrived on a smooth white beach to explore the mysterious village. Nineteen intricately carved totem poles stood proudly along the shoreline, monuments built by the natives who had apparently fled the community for some unknown reason. The tragic site inspired Kincaid to write a poem he titled "The Vanished Tribe."

> In a sheltered cove on a sunny strand
> An Indian village lay
> Where waving branches of the spruce,
> Cast shadows o'er the bay.
>
> The years passed on with cycling flight
> While chieftains rose and fell,
> The clan waxed mighty in its strength,
> And all its way seemed well.
>
> They reared their totems, burned their dead
> And hunted through the vales,
> Ensnared the salmon in the streams
> And ate the stranded whales.
>
> The white man came with reckless greed,
> And left his evil mark;
> The tribe sank down, but kept its way,
> And bore its tribal ark.

The fever rose with awful touch
And smote the redman sore
He murmured 'gainst the very gods
That haunt his wigwam door.

The Shaman failed to stem the tide
Of spirits ill and dread,
His rattle and his pompous words
Their anger simply fed.

The people fled with sickened hearts
To 'scape the awful doom,
And left the village as it stood
As silent as the tomb.

But "reckless greed" pervaded Kincaid's own party—his colleagues eyed the ornate totem poles with keen interest. Bleached by the sun and harsh winters, the poles, with their carved birds, bears, and contorted faces of strange sea creatures, fascinated the scientists. Here was a veritable windfall of souvenirs and collectibles to adorn the museums of their universities and scientific institutions. Deck hands from the *Elder* were called to shore to help dig up the old village's heavy monuments. While the workers labored in the unusual Alaska heat, the Harriman guests explored the remainder of the abandoned village, whose houses seemed to have been suddenly and mysteriously deserted. Children's clothing still hung from the decaying rafters of one of the wooden cabins and the scientists excitedly discovered bizarre religious masks, baskets, and carvings inside other houses. Ransacked boxes and scattered Indian artifacts pointed to the presence of earlier pillages, but, the Harriman scientists thought, the very best offerings of the village still remained in the imposing totem poles. Gnats flocked about the sweaty faces of the crewmen who strained against the obstinate heavy poles, and John Muir watched the difficult procedure in disgust, accustomed by now to the ways of his colleagues on the expedition. One of the poles proved doubly rewarding; when a nest of squirrels was discovered in the top of the pole, "two of the half-grown young

Gathering Souvenirs at Cape Fox (Averell Harriman in center, wearing white hat).

were caught and of course made into specimens."

The removal of the poles probably disturbed Muir more than the killing of the baby squirrels. It was not the first time the old naturalist had witnessed such a "robbery." On a trip to Alaska in 1879, he had protested when an archeologist in his party cut down a totem pole "with a view to taking it on East to enrich some museum or other. This sacrilege," he wrote later, "came near causing trouble." But causing trouble among the dignified Harriman company seemed incongruous, and the dismantling of the poles went on without a peep of protest, except from the overworked crew members—hauling totem poles off a beach and onto the ship was not ordinarily part of their jobs. Sympathetic to the workers, J. Stanley-Brown satirized the scientists in his poem "The Taking of the Totems."

'Twas in '99 and the day was bright,
  Cheerily, my friends, yo ho!
An abandoned village hove in sight,
And up comes a scientific witless might,
  Cheerily my friends, yo ho!

"Who'll go ashore today with me,
And gather some totems—say two or three?"
"Why, bless you, man, that's the job for 'we'!"
    Cheerily, my friends, yo ho!
    Cheerily, my friends, yo ho!
With a long, long pull and a strong, strong pull
    At the bottle to help us through
And we cheered without stint for the "science gent"
Who showed us the thing to do.

We launched the "napthas" which pulled like a Turk,
    Gleefully, my friends, yo ho!
With ropes and shovels we were soon at work,
And dug and tugged with never a shirk;
    Hopefully, my friends, yo ho!
We pulled and strained till the light grew shy,
Till our hands were blistered and our tongues were dry,
But those figures still grinned at the stars in the sky,

Displaying Trophy.

Souvenirs from the Deserted Village, Cape Fox.

Cheerfully, my friends, yo ho!
Cheerfully, my friends, yo ho!
With a long, long curse and a strong, strong curse,
    At the comrade who made us sore,
We slipped away to our ship in the bay,
    Vowing we'd work no more.

The "boss" knew nothing of Indian lore;
    Happily, my friends, yo ho!
He got very wet and mildly swore,
As in raiment scant he paced the shore
    "I'll have 'em yet, my friends, yo ho!
Just let 'pure science' have its fling,
Then I'll show them how to do this thing!"
And he sent for the crew who came with a swing
    Cheerfully, my friends, yo ho!
    Mockingly, my friends, yo ho!
With a long, long pull and a strong, strong pull
The trophies came down with a din
But we'll tell later on to our friends in town
How WE gathered those totems in.

While the scientists scavenged among the houses and the crew members struggled to bring the great poles down, the region's enormous black ravens watched the invasion from their perches on top of the poles, and their croaking protests echoed eerily across the water. Both Keeler and Elliot were too excited about the totem poles to notice the strange calls of the birds. Keeler's California Academy of Science and Elliot's Chicago Field Museum were each to receive two of the poles, and the two men proudly thought of the reception that would be forthcoming when the impressive totems arrived at their institutions. Harriman designated the University of Michigan to receive yet another pole, and the tycoon dragged out a pair of wooden carved bears that guarded the grave of one Indian for his own collection. Standing on their haunches, the ferocious-looking wooden bears left deep furrows in the sand as they were pulled toward the ship. The village's cemetery yielded yet another find for the scientists. Merriam discovered two Chilkoot blankets placed over a grave, and took one that seemed "in fair preservation."

Indeed, so numerous were the findings at the village, that the entire next day was spent hauling off its artifacts. In the unexpected heat, Dall took shelter on board the ship, Dellenbaugh leisurely sketched, and Fisher packed his numerous bird specimens, while the crew brought carving after carving back to the ship. At the close of the day, however, Harriman ordered the entire expedition back to shore so Curtis could take a group photograph in front of the ransacked village.

In the last few days of the long expedition, there was a sense of both celebration and denouement. After the photography session on shore, the group reboarded the ship for an evening of entertainment and homage to Harriman. Fernow and Gifford played their piano and violin before the assemblage. Afterward a nervous Fuertes sang a song for the group and Captain Kelly gave an animated rendition of a Sioux war dance. What a cynical Dall called a "love feast" followed the music and dancing. Speaker after speaker rose to commend Harriman for his great undertaking. A final recommendation was

Iliamna! Giant bold!
Snow-enshrouded to the Sea
Longthy graceful slopes will hold
The memory of the
H·A·E

Frederick Dellenbaugh, "The Memory of the H.A.E.," from "Chronicles and Souvenirs."

frivolously made to initiate a "Harriman Club" composed of expedition members, any two of whom "may have a meeting at any time" provided that each "pays for his own dinner." A riotous chanting of the familiar H. A. E. "war cry" rose up and Harriman gratefully acknowledged the appreciation of his guests, though he admitted that when the train of palace cars first left Grand Central Station in May "he did not realize the magnitude of the undertaking and the amount of work involved." But it was Captain Doran's night as well as Harriman's, and the ship's captain received his praises and accolades with hearty laughter as poems, cheers, and songs celebrated his leadership during the expedition. It was a night, wrote Muir, of "wild glee and abandon" before the tired merrymakers finally dispersed to their staterooms, echoing "Who are we? Who are we? We are, we are, the H. A. E.!"

CHAPTER THIRTEEN

"A LAPSE OF TIME AND A WORD OF EXPLANATION"*

When the *George W. Elder* sailed into Seattle early on the morning of July 30, its hold was packed with crates and boxes of birds, mollusks, fossils, Indian artifacts, plants, and mammals that would take years to sort and identify. Even the expeditioners' personal bags and satchels bulged with souvenirs and trinkets they had acquired on the long voyage. A flurry of press attention greeted the travelers as they returned to the United States, and newspaper coverage hailed the scientific significance of the expedition. In Portland, William Brewer told an impressed reporter from the *Oregonian* that the expedition achieved "far more" than he had originally hoped. The *World's Work* wrote a glowing editorial that heaped laurels on Harriman's wise use of money and his "admirable management and untiring effort." The New York *Daily Tribune* called the trip "an entire success," citing the discoveries of new glaciers and uncharted waters, the Eskimo studies, and the thousands of photographs taken.

Edward Curtis, uncredited for his photography in the news

*From *Ballads of a Bohemian* by Robert Service.

accounts of the expedition, unloaded his vast portfolio in Seattle. He hoped that his fellow passengers would order enough copies of his work to make the long voyage worth his absence from his own studio business at home. Even if orders for the photographs were not overwhelming, however, the trip to Alaska with Harriman would prove to be perhaps the most valuable Curtis had ever undertaken. His acquaintance, George Bird Grinnell, had become a close friend during the two months of the expedition, and Grinnell had invited him along on an upcoming trip to Montana to visit the Blackfeet Reservation. The mystical lure of the Indians' dying civilization spoke deeply to the photographer, and, within a month, he would find himself riding horseback with Grinnell across the high, windswept Montana valleys, befriending the Indians and seeing for the first time the still-sizable population of their tribes. The number of Washington Indians had dwindled, and Curtis would be amazed by the sight of thousands of Indians in Montana.

Thus Edward Curtis' lifelong devotion to preserving the culture and character of the American natives in photographs, on motion-picture film, and on phonograph recordings began as a result of the comradeship and experiences on the Harriman Expedition. After Curtis struggled on his own for seven years, J. Pierpont Morgan would become his benefactor, subsidizing his dedicated career traveling among the Indians.

Curtis and Gilbert both remained in Seattle, while the rest of the expedition journeyed down to Portland, Oregon, where the East Coast members would take a return train back to their homes.. But Merriam and Dall had more exciting things on their minds than returning to the stuffy halls of the Smithsonian. After a reluctant farewell to their fellow expeditioners, the two met their old friend Gifford Pinchot and headed by train to San Francisco, following their companions Muir, Keeler, and Ritter, who had left immediately to rejoin their families in the Bay Area. Dall was headed for Honolulu, where he would be appointed Honorary Curator of the Bishop Museum, a position he would hold for the next sixteen years. But the Smithsonian would continue to be Dall's workplace until his death in 1927. There, surrounded by his immense library and vast paleontological collections, he wrote hundreds of treatises on mollusks and on Alaska, confirming his national reputation as the country's expert on the region. Shortly after the expedition, when Frederick Dellenbaugh wrote Dall about donations for a silver punchbowl that the expedition members had proposed to give Harriman, the old scientist bristled at the notion of degrading science in such a manner. It was, he scribbled to Dellenbaugh, a "philistine idea" and a "mistake" to make such a gesture. Science remained a realm of eminent seriousness for Dall, and the silver punchbowl seemed to denigrate the work he held in such high regard.

When Dall parted company with his companion Merriam, the latter was headed to San Francisco with Pinchot; he would stay in California for nearly three months. Merriam was worried about his appointment by Harriman as editor of the Committee on Publication. It would mean, he feared, "a large amount of additional . . . work." But the biologist had no inkling that the project would consume the next twelve years of his life. When Harriman died in 1909, he left a fund for Merriam to use in his research and, thus subsidized, the scientist continued his work and solidified his reputation as a pioneering mammalogist.

California was a welcome sight to Muir and Keeler. They could well understand how Merriam could plan to spend an additional three months in the balmy Bay Area. On his return to his home base, Muir immediately picked up his work in wilderness preservation, and, in the ensuing years, he became increasingly recognized as a leading force in the creation of national parks and the passage of national conservation acts. In 1903, he camped with President Theodore Roosevelt at Yosemite, an outing which convinced the President of the need for wilderness preservation. The naturalist also kept up an active contact with Harriman, who encouraged Muir to write and to maintain his political lobbying efforts. When Yosemite Valley was included in Yosemite National Park in 1906, Har-

riman used his power behind the scenes in Congress to rally
support for his old friend's cause. Harriman's death in 1909
inspired Muir to write a slim volume on the independent, determined businessman, extolling his virtues of loyalty and generosity. "Of all the great builders—the famous doers of things
in this busy world," Muir wrote, "none that I know of more
ably and manfully did his appointed work than my friend Edward Henry Harriman." Though today the name of John Muir
connotes images of the lonely wilderness devotee whose solitary hikes in the Sierra and Alaska have become legendary to
new generations of nature lovers, he was also no stranger to
high finance and the Congressional lobby—especially in a good
cause.

Reunited with his wife and daughter, Charles Keeler remained in San Francisco, writing poetry and an occasional book
of prose while his friend Muir roamed the mountains. Keeler
would go on to become the "guiding spirit of the simple life
philosophy," perhaps best exemplified by his architectural
classic *The Simple Home*, which inspired the work of Bernard
Maybeck and a first generation of original California archi-
tects. In 1906, Keeler survived the great San Francisco earth-
quake and wrote dramatic accounts of the event.

When John Burroughs arrived back in New York State, his
star steadily rose in the public eye until he became a sort of
"sage and prophet" to the numerous visitors who trekked to
his rustic cabin, called "Slabsides." Among his friends Burroughs counted Thomas Edison and Henry Ford. His friendship with Theodore Roosevelt led to their famous joint
campaign in 1903 agains the "nature fakers," who, they
charged, distorted their experiences with nature and the wilderness. When Burroughs died in 1921, four days before his
eighty-fourth birthday, he was a beloved public figure. After
his death, however, his romantic, sentimental writings were
forgotten in the modern age ushered in by the 1920s.

Young Louis Fuertes returned to the East Coast eager to
present his bird portfolio to his teacher, Elliott Coues, who

would use the rich material to illustrate the latest edition of the *Key to North American Birds*. But, within only a few months, Coues would be dead, and Fuertes would continue his work without the encouragement and support of his old friend and mentor. The young naturalist's career, however, was already set in a direction of which Coues would have been proud. He traveled widely during the upcoming years to study the birds of strange lands, and as his paintings achieved wider and wider acclaim, he worked steadily at his studio in Ithaca, New York, producing the paintings of rare specimens of birds for which he became famous. A year before his untimely death in 1927, Fuertes traveled to Abyssinia with Wilfred H. Osgood, the same biologist whom Merriam had spotted climbing the slopes of the White Pass Ridge in Alaska twenty-eight years earlier on the Harriman Expedition.

Dellenbaugh's adventures in exploration were far from over in 1899. In the next several years, he would travel in Iceland, Norway, the West Indies, and South America, but he was always drawn back to the comfortable smoking rooms of the Century Club, the Explorer's Club, and the Cosmos Club. Wreathed by the smoke of a good cigar, he entertained his friends with lively tales of his journeys. He so regaled his Century Club companions with Harriman Expedition stories that one of his listeners wrote "I feel as if I were talking to . . . [Commander] Peary!" Dellenbaugh carefully recorded his travels both in drawings and in his diaries, in addition to his animated oral commentaries. In 1908, he published the noteworthy *A Canyon Voyage*, recounting his adventures with the Powell Expedition in 1871. Up to his death at eighty-two in 1935, the artist and explorer continued to paint and write popular books. He was, perhaps, the Lowell Thomas of his day.

Gifford also kept up a lively pace in upper-class circles on the East Coast until he died six years after the Harriman Expedition. Today the artist is best known for his illustrations of Theodore Roosevelt's books on the West. Charles Palache, of course, eagerly returned to his patient fiancée in the East; their wedding took place soon thereafter. Gilbert continued his illustrious career with the U.S. Geological Survey. As a result of the Harriman Expedition, he brought Alaska into his incredible geological studies of the American West, and before he was finished, Gilbert, suffering ridicule at the hands of colleagues, became one of the first Americans to study the geology of the moon. Even more significantly, his engineering-mechanics approach to geology laid the groundwork for and foreshadowed the plate-tectonics revolution. As for the genial Captain Doran, he continued sailing luxury liners up and down the Pacific Coast until 1907, when a shipwreck claimed his life. His steamship *The Columbia*, ironically equipped with "four water-tight bulkheads," sank after colliding with another ship off the California coast at Mendocino.

Edward Harriman descended from Alaska into a world of national controversy over his tremendous new power. The expedition he had masterminded helped to legitimatize the great accumulation of wealth by the few during this era of corporate consolidation and industrial expansion by waving a banner of social responsibility from the highest rungs of the corporate ladder. But the favorable publicity created by his Arctic adventure was short-lived. In the next ten years, he earned international notoriety with his numerous acquisitions and bold assertions of power, and public opinion turned against the tycoon, who conducted his business with little thought of his reputation. Indeed, he never bothered to publicize his occasional philanthropic gestures, shunning the applause that men like Carnegie and Morgan heaped upon themselves with their well-advertised libraries and museums.

In later years, the public would frown on Harriman's increasingly grandiose business schemes. Six years after the Alaska expedition, Harriman traveled to Japan with the ostensible purpose of promoting United States commerce in Asia. But beneath this goodwill tour, according to one historian, lay "a plan for a round-the-world transportation line, under unified American control, by way of Japan, Manchuria, Siberia, European Russia and the Atlantic Ocean." The connection across the Bering Strait was crucial to the plan—such a link had, in fact, been considered by the tiny railroad that Harriman visited in 1899. Its ambitious owners had long-range plans of constructing "an all-land route to Europe under Bering Strait." One can only speculate whether the small railroad had stirred Harriman's imagination in 1899 or whether he traveled to Alaska to investigate the feasibility of a rail route to Russia. Regardless of its source, however, Harriman's dream was never realized, of course. Unable ever to rest from his obsessive business negotiations, Harriman drove himself relentlessly until his death in 1909.

The ten years since his grand Alaska voyage had seen his reputation among the press change from that of a benevolent philanthropist to one of an arrogant, power-hungry tycoon. Coming as it did at a turning point in Harriman's career, the expedition was a lavish public gesture that won its leader immediate national acclaim and recognition. From the scientific venture, Harriman learned the value that such undertakings could hold for public relations, but he would not make many such gestures during his career, even though he had obtained from the expedition an elite group of "public friends" who would prove to be worth a legion of paid political hacks. John Muir publicized Harriman's contributions "in developing the country, and laying broad and deep the foundations of prosperity." Such praise meant far more to the railroad baron than did widespread public esteem. Indeed, he even seemed to hold the public in disregard throughout his political maneuvering.

Harriman's young son on the expedition, eight-year-old Averell, was destined to make his own mark on politics. After following his father into the railroad business, the younger Harriman found the world of government and diplomacy more to his liking. He was appointed Ambassador to the Soviet Union and, later, to Great Britain in the 1940s. As a Democrat, he was elected Governor of New York in 1954, serving one term before he was defeated by one of his colleagues in the powerful reaches of the American upper class, Nelson Rockefeller. Harriman and Rockefeller moved easily from elite corporate circles into governmental positions. As the last surviving member of the expedition, Averell Harriman epitomized the adventuresome nineteenth-century company that had journeyed together to Alaska in 1899.

As for the grand dreams of a railroad around the world and the tunnel under the Bering Sea, they were lost amid the mighty political convulsions that characterized a new age of global divisiveness—the twentieth century.

EPILOGUE:
MR. HARRIMAN'S
LEGACY

CHAPTER FOURTEEN

EDWARD S. CURTIS' ALASKAN VISION

One of the principal reasons for remembering the Harriman Expedition is that it changed the photographer Edward Curtis' life and led almost directly to the production of one of the great American masterpieces, Curtis' twenty-volume work *The North American Indian*, which was accompanied by twenty portfolios of photographs of eighty different tribes.

On the Harriman Expedition, however, Curtis' mission was not primarily to photograph Indians, but rather to document the voyage, both in posing the expedition members and in scientifically capturing typical Alaskan scenery. He was recruited by C. Hart Merriam owing to an interesting circumstance. One winter day in 1897, Merriam, Grinnell, and Gifford Pinchot found themselves in perilous straits high up on Mount Rainier, and a local society photographer, out for a climb with his female assistant, came to their rescue. The photographer was Edward S. Curtis. After they climbed down from the mountain, Curtis entertained his new friends by showing them literally hundreds of photographs of Mount Rainier and the Puget Sound country that he had taken as a hobby. When the time came to select a photographer for Mr.

Harriman's expedition, both Merriam and Grinnell remembered Curtis. The latter accepted Merriam's invitation with alacrity and brought along an invaluable assistant, D. J. Inverarity. The two men made photographs on the expedition that even today are indistinguishable as to which photographer made which picture. Inverarity, however, dropped from photographic history and became a lawyer, while Curtis, inspired by Grinnell, went on to his life's work on the American Indian.

As the previous illustrations in this book indicate, Curtis and Inverarity were not the only people on the expedition who took photographs. Both Merriam and Grove Karl Gilbert proved to be accomplished photographers, but, in catching the excitement of the expedition, Harriman perhaps outdid them all. In 1899 the camera was still something of a novelty, especially since dry plate and even paper film had recently made it relatively easy for anyone to take a picture. The age of the Kodak had begun, as had the era of the personal tourist photograph, and Alaska at the time was literally being plundered by photographers, until in a very short time its scenic wonders became clichés.

Curtis, however, brought something more to his photos, something subtle and at times difficult to understand. To those who are familiar with his haunting, romantic North American Indian photographs, his pictures of the Harriman Expedition do not look like the work of the same photographer at all. For one thing, on the Alaskan expedition he photographed few Indians and produced none of his soon-to-be-memorable portraits. When he did photograph the natives, he saw them as part of a bleak landscape invaded and polluted by the white man. If one looks closely at his native scenes, one sees the Indian and Eskimo in the sad process of acculturation. In depicting their dilapidated dwellings, their slumlike sealing camps, as well as their deserted villages, Curtis slyly presented a cultural message. But over all these genre scenes hangs an air of unreality, as if the red men and their dwellings were

Columbia Glacier from Heather Island, by E. S. Curtis.

apparitions. They appear as ghosts in a remote surreal landscape.

When he turned specifically to the landscape, which included the sea and floating icebergs as well, Curtis was faced with several different problems. One was that of the immense scale of his subjects. Mile-long glaciers, towering mountains, long, lonely inlets and river valleys offered the photographer, armed with a six-by-eight-inch camera, a special challenge. He could not paint large canvases like the famous iceberg artist Frederick Church. Instead he had to capture panorama, scale, texture, scientific detail, and mood all in a comparatively small frame. In executing this task, his previous experience on Mount Rainier served him well. Whenever he could, he composed a scene in his mind, often with reference to visual romantic prototypes, such as the engravings of Doré, before he shot it. His trademark was usually a dark, richly textured foreground succeeded by bands of light and dark receding into the background, where the Alaskan sky played a dramatic part. Always conscious of light, especially the strange Arctic light, Curtis used it to highlight the texture he so much prized in his photographs. Whether it was a rocky, sedimentary beach, an undulating glacial surface, or even the fluted mass of the glacial face itself, Curtis caught the texture. He also caught fields of Arctic flowers in the same way; on one occasion he took a close-up view of tangled reindeer moss and huckleberry in a manner similar to the mysterious calligraphic depths later reached by Jackson Pollock.

At the same time, Curtis was always conscious of his duty to provide scientific documentation of glacial action. He often went out to the glaciers with the geologist Grove Karl Gilbert, who was seeking to understand the whole mechanism of glaciation. Curtis' photos of glacial surface, glacial recession, glacial "attacks" on whole forests, the rubble-strewn aftermath of glacial recession, and newly made icebergs all served a serious scientific purpose. They documented the advance or retreat of glaciers in the year 1899 and thus could be compared with photos from other years, and they also illustrated the profound processes in which Gilbert was so very interested.

And yet Curtis served these documentary purposes in such a way as to produce distinctive photos of genuine artistry. A master of luminist techniques, Curtis rang the changes from the softness of quiet grandeur to the menacing immediacy of potential· storm and disaster. In this he duplicated with the camera virtually every effect achieved by the distinctive American luminist painters of the nineteenth century. The strange Arctic light made such effects too obvious to pass up for any ambitious artistic photographer, and Curtis had a natural eye for the uses of light. In one classic photograph he captured the *George W. Elder* and a stately three-master afloat like toy ships in the serene vastness of Orca Harbor at eleven p.m. It was still daylight.

In general, the Alaskan experience was so overwhelming to Curtis that his photographs took on a surreal aspect. His forte was the immense empty landscape in which people or settlements, if one found any, seemed like a temporary intrusion.

The Barry, by E. S. Curtis.

The Muir—From the West, by E. S. Curtis.

The Way to Nunatak—Ridged Ice, by E. S. Curtis.

From the Strand Near Muir House, by E. S. Curtis.

The Harvard, by E. S. Curtis.

Orca Harbour at 11 p.m., by E. S. Curtis.

House and Hearth—Plover Bay, Siberia, by E. S. Curtis.

Sealers' Camp—Yakutat Bay, by E. S. Curtis.

Eskimo Village—Plover Bay, Siberia, by E. S. Curtis.

A Sealer's Shack—Glacier Bay, by E. S. Curtis.

In Muir Inlet, by E. S. Curtis.

Last View of the Pacific, by Edward S. Curtis.

The Resurrected Forest—Near Muir Glacier, by E. S. Curtis.

Whether it was a strange Russian church, a Chuchchi skin tent, or an oddly antediluvian whalebone "hearth," Curtis' subjects were always seen against the mysterious sublimity of nature. His imagination, like that of most of his fellow-voyagers, was essentially transcendental, but his work was also strongly tinged with a sense of the dreamlike quality of nature. Alaska was, after all, not like cozy New England; it was not even like spectacular Puget Sound. It had a vast expressionistic quality of its own. Everyone on the expedition noted this, and Curtis caught it with his cameras as he reached for something beyond the surface reality that might have appealed to the journalist or the tourist.

Harriman understood this. When he decided to present everyone on the expedition with a "souvenir album," he chose to make it exclusively an album of photographs. Curtis was left to see to its composition. All during the spring and summer of 1900 he worked furiously in his Seattle studio, developing and printing photographs for the *Souvenir Album* as well as for the first two volumes of the expedition's official *Report*. He submitted his work to C. Hart Merriam in Washington, who proved to be a hard taskmaster. Merriam frequently sent the photos back because Curtis had cropped them poorly or because he failed to get the waterline or horizon lines exactly horizontal. Merriam's eye and tastes thus also came into play in the selection of Curtis photographs destined for both the *Souvenir Album* and the published *Reports*. Almost as much as Curtis, he was responsible for the pictorial planarity that, together with the manipulation of light, characterizes Curtis' final luminist photographs.

Meanwhile, with the negatives in his possession, Curtis was printing and selling Harriman Expedition photos at his studio in Seattle. Though his business was reasonably lucrative— enough so to get him started on his North American Indian project—few of these Curtis Studio photographs have survived. The originals can be seen only in the extremely rare *Souvenir Album*, a selection from which is presented here.

Today these photographs have a dual interest. They indicate something of the beginnings of Curtis' magnificent career—a beginning that seems, superficially, strangely different from his later work, which was exclusively taken up with Indians. To the casual viewer they also appear merely as pictures of now-familiar scenery; but viewed closely, their romantic, transcendental, even surreal quality foreshadows the way in which, as "The Shadow-Catcher," Curtis would portray the American Indian. He placed both Alaska and the Indian somehow beyond the reach of contemporary civilization in a timeless realm. This was his legacy to the Harriman Expedition.

It was not his only legacy, however. In a sense more immediate to our story, he reflected with precision the emotional and intellectual attitudes of the expeditioners as they explored Alaska. Like his more elite comrades, Curtis looked past the transient pillaging of Alaska to that which he deemed permanent—the silent, everlasting power of nature.

CHAPTER FIFTEEN

SCIENTIFIC RESULTS

It should be apparent that the Harriman Expedition was more than a rich man's junket—a summer vacation in Alaska's wonderland. It was also more than a grandiose railroad-building scheme. Clearly Harriman intended that his boatload of "scientifics" would make a genuine contribution to science, and he spared nothing to make this possible. The *Elder* was entirely at the disposal of the scientific corps and sailed to the sites that they deemed important. When the voyage was over, Harriman subsidized the publication of thirteen sizable volumes that constituted a report on the results of his expedition. C. Hart Merriam devoted twelve years of his life to directing the production of the volumes and seeing to it that the large collections brought back by the expedition were placed in the right institutions and in the hands of the right scholars. As a result, the production of the Harriman Expedition *Reports* became a large scientific enterprise that carried far beyond the work of the original twenty-five savants who sailed aboard the *Elder*. At least forty-six scientists and numerous assistants worked on the reports at locations as various as Washington, D.C., Ithaca, New York, and Berkeley, California. In this sense, a modern, professional team concept, much like that employed by Clarence King and John Wesley Powell in the U.S. Geological Survey, was brought to bear in studying and working up the expedition's results. It was no small operation.

Nonetheless all the scientists involved recognized that the expedition was a reconnaissance rather than a comprehensive or definitive survey of the region. The quality of the scientists aboard the *Elder* was much too high for them to be so naïve as to claim anything else. They well knew that numerous previous expeditions had probed the area, going back to Captain Cook and Vitus Bering. William Healey Dall, himself a previous explorer of the territory, made this clear in an essay on the discovery and exploration of Alaska included in the second volume of the series. Virtually all the scientific papers that made up the expedition *Reports* mentioned previous studies of the region, some of them very recent, and the scientists were well aware of the comprehensive surveys, on both land and sea, then being conducted by the U.S. Coast and Geodetic Survey, the Fish Commission, and the U.S. Geological Survey, which had turned its attention to the vast interior of Alaska. Without hesitation Harriman's scientists made use of materials from all the other expeditions when they proved helpful in illustrating a point, providing new data, or making a survey of regional species more complete. The geologist Grove Karl Gilbert put the nature of the expedition most clearly when he declared at the onset of the geological report: "The geologic field work of the expedition may properly be dramatized as a reconnaissance; and in this respect it resembles the greater part of the work which had previously been accomplished in the same region. While it was in progress there was much more active exploration in the interior of the Territory, chiefly by members of the United States Geological Survey, and that work has continued in later years."

Elsewhere, in his introduction to the report on glaciers, he carefully delineated the scope of his work: "the growth of knowledge of Alaska glaciers is so rapid that a summary of existing knowledge would have but transient value. . . . It has seemed best, therefore, to make the present report primar-

ily a record of the data gathered by the Harriman Expedition.
. . ." Gilbert, like his scientific colleagues, merely wished to
place his accurate observations in the growing context of scientific knowledge. It is this professional restraint that, as much
as anything, enhances the quality of the expedition *Reports.*

After the publication of the first two volumes of *Reports* in
1901 and 1902, the succeeding *Reports* became increasingly
specialized and esoteric. For the most part they were written
by scientists for other scientists, and were intended as a reference survey and a guide to collections and resources. Volumes I and II, however, seem directed at the general public,
and, as such, they were well received. The first volume fea-
tured the most popular topics in sections written by the best-
known authors. The famous naturalist John Burroughs wrote
the official narrative of the voyage, replete with anecdotes and
observations of natural and human curiosities. John Muir,
though he tried to be scientific, wrote what amounted to an
appreciation of the glacial wonders of Alaska. He stood in awe
of "the restless motion of those mighty crystal rivers . . .
ceaselessly flowing and grinding, making soil and completing
the sculpture of their basins." He enjoyed the "glorious views
of Malaspina's crystal prairie," fell into wonder at whole "new
islands" being added to Alaska's famous volcanic archipelagoes, and stood transfixed when "the heavens opened and
[Mount] St. Elias, gloriously arrayed, bade us welcome, while
the heaving, plunging [ice]bergs roared and thundered." At the
conclusion of his essay, which consisted largely of observations about the advance and recession of glaciers he had observed, Muir had one of those transcendental epiphanies so
dear to the heart of every Ruskinian: "The sail down the coast
from St. Elias along the magnificent Fairweather Range, when
every mountain stood transfigured in divine light, was the
crowning grace and glory of the trip and must be immortal in
the remembrance of every soul of us."

While Muir was flirting with the sublime, however, Grinnell was looking in the other direction. His essay on the Indians of the region was, he realized, compared to the mighty

labors of Franz Boas, a mere cursory glance at the subject. But he went through all the obvious banalities, including a description of the Kwakiutl potlatch ritual, in order to get across his social message. "The outlook for the immediate future of these Eskimos is gloomy," he wrote. Then he amplified his remark: "The rush to the Coast gold fields has brought to them a horde of miners who, thinking only of themselves, are devoid of all feeling for others of their kind. . . . White men uncontrolled and uncontrollable already swarm over the Alaska coast and are overwhelming the Eskimo. They have taken away their women, and debauched their men with liquor; they have brought them strange new diseases that they never knew before, and in a very short time they will ruin and disperse the wholesome, hearty, merry people whom we saw at Port Clarence and Plover Bay."

George Catlin, at the height of the Romantic Era was scarcely more direct, though more eloquent. Elsewhere Grinnell, in painting a devastating picture of the rapaciousness of the salmon industry, pointed out that the ruthless and wasteful white capitalists were usurping a domain of property long held sacred by the Indians of Alaska. "For hundreds of years," he wrote, "the Indians and the Aleuts had held these fisheries . . . with an actual ownership which was acknowledged by all and was never encroached upon. . . . No Indian would fish in a stream not his own." In both his essays—on the Indian and on the salmon fisheries—Grinnell lamented the invasion of the competitive and rapacious white man, but then in an odd, fatalistic turn, he concluded in the spirit of the Spencerian anthropologist Lewis Henry Morgan, "there is an inevitable conflict between civilization and savagery, and wherever the two touch each other, the weaker people must be destroyed." It is ultimately this tragic sense of the red man's fate that so informed the poignant photographs in his friend E. S. Curtis' *The Indians of North America*.

Volume II of the *Reports* continued the popularized survey of the expedition's findings. Each author of an essay struggled to be scientific, but invariably what resulted was appreciation, if not pure impression. Dall's "The Discovery and Exploration of Alaska" was an all-too-brief survey of Alaska's early history. Charles Keeler's "Days Among Alaska's Birds" was exactly what the title suggested—a bird-watcher's hymn, illustrated by Louis Agassiz Fuertes' wonderful drawings in color. Fuertes, though only a youth and on his first important field assignment, made such excellent use of his talents in drawing the comic horned and tufted Puffins, long-tailed Jaegers flocking through the sky, the mottled Harlequin Duck, the violet-green and white-crested screaming Cormorant, and dozens of other birds, that it was evident he was destined to become the twentieth century's John James Audubon.

Still, in Volume II the German-educated B. E. Fernow surveyed "The Forests of Alaska" with as much precision and order as he could muster from a brief reconnaissance. He concluded that Alaska would never be a great source of timber. The wood was inferior and the conditions for lumbering much too difficult. "At present," he concluded, "it pays to import lumber from the Puget Sound country or other points of the lower coast." At least he had temporarily discouraged timber prospectors.

Henry Gannett wrote a *very* general survey of Alaska's regions in his "General Geography," then concentrated on the territory's mineral resources—gold, copper, and coal—concluding that they were abundant, and thus inadvertently encouraging even more prospectors of the sort that had already debauched the Indians and infested the ramshackle waterfront towns all over Alaska's coasts, or else perished up some frozen river in the interior. Gannett, however, saw Alaska in a different way. He remembered Yellowstone and Yosemite, asserting, "For the one Yosemite of California Alaska has hundreds." In his view, Alaska's future was bound up in its magnificent scenery. Linking the transcendental and the commercial he prophesied: "The Alaska coast is to become the show-place of the earth, and pilgrims, not only from the United States, but from far beyond the seas, will throng in endless procession to see it. Its grandeur is more valuable than the gold or the fish or the timber, for it will never be exhausted. This value, measured by direct returns in money received from tourists, will be enormous; measured by health and pleasure it will be incalculable." Gannett, like his contemporary the philosopher of pragmatism William James, knew "the cash value of an idea."

Except for William H. Brewer's rather inconsequential speculations on "The Alaska Atmosphere," much of Volume II of the Harriman *Reports* was taken up with dissertations on the magnificence of Alaska's natural wonders, including its wildlife, aboriginal inhabitants, and exploitable resources. But a contrapuntal theme runs through most of the essays, even Dall's history: The advent of the white man had spoiled, and would continue to spoil the holy wilderness. Certainly Grinnell's lurid description of the salmon industry, with its ruthless cutthroat competition and its appalling waste, together with M. L. Washburn's unintentionally ironic little dissertation on "Fox Farming in Alaska," contributed mightily to the overall feeling of perplexity that infuses the volume. To read it is like reading a reprise of the history of the trans-Mississippi frontier, wherein the inevitable juggernaut of civilization rolls over unspoiled nature, and the confused citizens of the republic must pause in their moment of triumph to lament the tragedy.

Beginning with Volume III, *Glaciers and Glaciation*, the *Reports* turn strictly professional. Volume IV, *Geology and Paleontology*, was the combined work of B. K. Emerson, Charles Palache, William Healey Dall, E. O. Ulrich, and F. H. Knowlton. Emerson generally surveyed the structural scene; Palache concentrated on minerals and mining, casting an admiring eye on the Treadwell Mine; Dall, Ulrich, and Knowlton focused on paleontology. In Gilbert's view, as chairman of the Committee on Geology, the geologists were still limited to reconnaissance. The most important findings were in the fields of glaciology and paleontology. He was most impressed by the location and correlation of fossil evidence "of slates and shales in three widely separate localities—Yakutat Bay, Prince William Sound, and Kodiak Island—and the determination of their

age as Jurassic." This formation, broadly distributed over the territory, was "the dominant constitution of mountain masses which have a long history, including base-leveling and subsequent uplift and dissection."

E. O. Ulrich was not certain that this was correct, insofar as he believed the critical horizon might be either Eocene or Jurassic. But Gilbert read the evidence as placing the mountain formation in the Jurassic and the lowland setting for the volcanic action in Alaska's archipelagoes as Eocene. In fact he pointed out as one of the expedition's triumphs the discovery of Eocene molluscan fauna for the first time in Alaska. This indicated that "the region was already the scene of volcanic activity in early Tertiary time." No one, however, really studied Alaska's volcanic activity beyond Merriam's historical and descriptive essay on the evolution of Bogoslof's two volcanic islands. At a time when Von Richtofen's classic on the sequence of volcanic eruption had recently become prominent, and when Clarence Dutton had published widely on mountain-making, Pacific volcanoes, and the "isostacy" principle, no member of the expedition saw fit to inquire into volcanoes, earthquakes, or coastal uplift and its causes, which were as characteristic of Alaska as its glaciers. Indeed they were crucial to any understanding of Alaskan geology. And, if they had been studied by Gilbert with the shrewdness and insight that he brought to the study of glaciers, the theory of plate tectonics might have been born on the Harriman Expedition to the active North Pacific Rim.

The year 1904 saw a perfect explosion of Harriman *Reports*. These included—in addition to Volume III, *Glaciers and Glaciation*, and Volume IV, *Geology and Paleontology*—Volume V, *Cryptogamic Botany*, Volume VIII, *Insects, pt. I*, Volume IX, *Insects, pt. II*, Volume X, *Crustaceans*, Volume XI, *Nemertians and Bryozoans*, and Volume XII, *Enchytraeids and Tribicolous Annelids*. Merriam's scientific assembly line was in high gear, though Volumes VI and VII mysteriously failed to appear and have not yet appeared to this day, possibly because they were slated to describe Alaska's mammals through the pen of Merriam himself, and the great taskmaster never got around to the assignment. The project rolled on, however, and in 1905 Dall and C. C. Nutting completed Volume XIII, *Land and Fresh Water Mollusks and Hydroids*. In 1909 E. H. Harriman died. Between 1905 and the time of his death no new volume had come out, suggesting perhaps that a healthy Harriman had been the driving force behind the scientific assembly line though its headquarters was in the Smithsonian. After Harriman died, the Smithsonian took over publication of the *Reports*, and in 1910 reissued them all with new title pages. Belatedly, in 1914, the last volume was published, Addison Emery Verrill's life work on *The Shallow-water Starfishes of the North Pacific Coast from the Arctic Ocean to California*. Mrs. E. H. Harriman financed its publication in two volumes, Part I being text and Part II plates, thus, except for the missing Volumes VI and VII, completing one of the most lavish publications of the age. Any suspicion that Harriman or his family were less than sincere in their support of science must be removed in the light of their persevering support for the long series of very specialized volumes.

For the most part, the biological volumes amounted to compilations and lists of fascinating creatures that floated, swam, flew, or stung. The volume on crustaceans should be of special interest to gourmets with a partiality to shrimp and crabs— Harriman's researchers had discovered hundreds of new species. In fact, the chief contribution of the biological volumes is threefold: in them the scientists brought to light several hundred new species and over fifty new genera; they charted the geographic distribution of these species and genera, with Dall outlining the entire Arctic habitat of mollusks; and they conscientiously dissected, scrutinized, characterized and listed all the species in every category. In each volume the Harriman Expedition collections were supplemented by dozens of other collections that accrued during the time the works were in progress. This made the volumes the most up-to-date, definitive works on their subjects and standard reference works for all subsequent generations of researchers.

Virtually none of the biological volumes ventured into the realm of theoretical speculation except that of Addison Emery Verrill of Yale, who was Merriam's old teacher. In studying starfishes, he found a long evolutionary sequence from the primitive Devonian five-pointed starfish to extremely complex starfish with as many as forty "points," complete with eyes, motor mechanisms, and suction devices that enabled them to cling to rocks. Verrill had a host of questions that grew from his research. For example, what was the evolutionary mechanism that spawned the forty-pointed starfish, and what explained the fact that forty-pointers and five-pointers lived simultaneously in the present day along with all the variants in between? What in particular did this say about Darwin's hypotheses? He was equally fascinated with starfish sensory mechanisms, nervous systems, and choice of locomotive direction. Why and how did a starfish move in one direction or another? Did it not have a head and tail that tended to turn locomotion in a predominant direction? As any student of space travel should be aware, this was not idle speculation.

Perhaps the most profound volume in the whole Harriman series was Grove Karl Gilbert's work, *Glaciers and Glaciation*. Gilbert had a peculiar angle of vision on science in general and geology in particular. As Stephen Pyne has written, in an age of historically oriented geology aimed at developing holistic theories, Gilbert was a throwback to classicism and the technique of distinct classical experiments. More often than not, he eschewed the historical approach to earth evolution in favor of what he could learn from Archimedes. Well aware, as he had declared in "The Inculcation of Scientific Method by Example," his presidential address to the American Society of Naturalists in 1885, of the importance of personal point of view and hence personal "style" in science, Gilbert brought the mind of an engineer to the study of geology. Long before Heisenberg, he knew that the observer in science affected the outcome of the experiment. A case in point: in the 1870s, while Wheeler Survey geologists were structurally and stratigraphically mapping the whole West, Gilbert concentrated upon the mechanics of the laccolite or dome mountain as exemplified by the Henry Mountains. By concentrating on that one range and its mechanics, he was able to adduce a theory of very broad application. As Gilbert put it in his Alaskan report, "When an observer views a complex phenomenon his attention is naturally directed to the peculiar features which his previous training enables him to appreciate—he 'sees what he has eyes to see'; and the difference of eyes makes the work of independently trained observers more or less complementary."

In his role as "glaciologist" on the Harriman Expedition, Gilbert knew that his work could not be definitive. His was, as he said, "a reconnaissance." Characteristically "seeing what he had eyes to see," he concentrated upon making the most careful empirical studies that he could of each glacier, employing maps and thousands of photographs—his own, Merriam's, Curtis', which he deemed the best, and any others he could obtain—to add to the detail of his very localized studies of each glacier. His intent was to build a body of data and observations that were so reliable for their place and time as to be of use to others that followed him.

Gilbert divided *Glaciers and Glaciation* into three long chapters: "Existing Glaciers," "Pleistocene Glaciation," and "General Considerations as to Glaciers." Though the sections seem to break down neatly into present conditions, historical geology, and general principles, Gilbert's classical mechanistic intellect ran through each part. Moreover, he brought to the study of glaciers eyes conditioned to the alluvial fans of the Great Basin, Rocky Mountain basalt flows, and the rushing streambeds of the Colorado River system. To him glaciers were simply another form of hydrology whose characteristics had always been a primary focus of his work. In Alaska, as Stephen Pyne has pointed out, Gilbert concentrated on three large topics: "glacial climate, glacial topography, and glacial motion."

In the first instance he suggested that changes in the temperature of the ocean rather than of the air masses accounted for general changes in the climate that affected Alaska. But he refused to be trapped by a general theory. Though broad climatic changes had occurred, each glacier was affected in a unique way, so that overall cooling or warming trends might have opposite, unexpected effects on individual glaciers. As he put it, "the combination of a climatic change of a general character with local conditions of varied character, may result in local glacier variations which are not only unequal but opposite." In a sense, glacial change depended in part upon the particular topographical setting of the glacier, say to the lee or windward side of the mountain. Thus the behavior of glaciers involved a great many complex variables and was not subject to one overarching theory. These variables were in constant tension, and hence in an equilibrium revealed at specific times and places only to empirical observers.

Glacial topography fascinated him. He observed that the surface of glaciers appeared to be smooth, worn down by wind and other forms of erosion, but the underside of the glacier was anything but smooth. It carried along with it all the det-

Geological Map of Alaska by B. K. Emerson.

ritus that lay in its path—boulders, sand, tree trunks, even water. This caused friction as the glacier pressed upon the earth, and made the glacier behave differently from a stream bed. In streams debris was picked up and dropped at periodic intervals; the glacier, owing to its comparative viscosity, carried everything along with it. It also embodied "shear," which meant that it moved forward at uneven levels, sometimes pulling against itself, so that the bottom might be stuck on a resistant rock structure while the top layer slid forward more easily.

In the field, Gilbert studied the track of the glacier, noting that it carved out valleys in two ways, by grinding and "plucking." "Plucking" referred to picking large boulders from their matrix and pushing them along in the glacier, leaving holes or "kettles." Gilbert came to believe that "plucking" was perhaps a more significant sculpting factor than grinding, though the multiple factors of velocity, pressure, the rock abraded, and the quality and quantity of abraded particles all governed earth-sculpting action to some extent. Details like these fascinated him and he missed very little. The growth of forest in relation to a glacier told him whether the glacier had advanced or retreated recently. A "hanging glacier," or one left high above the trunk glacier like a frozen waterfall disconnected from the main stream, told him something about the comparative velocity of the glaciers, trunk and branch, and about the sequence of their creation.

Ultimately, Gilbert concluded, the main factor in glacial motion (the trend of his questions) was velocity. Velocity was determined by gravity in the main, but also by internal shear, rock resistance, and friction. For Gilbert there was always multiple causation. In theory these multiple causes could be measured, but in fact, of course, it was difficult if not impossible (or even fatal) to get inside or beneath a glacier. The best course was to study its wake in eroded valleys, moraines, and rutted plains. This constituted a paradox: the empirical Gilbert was invariably forced back to abstractions and models in describing phenomena or conceptualizing hypotheses. This took him far from the descriptive aspects of geology and into the "non-common-sense" realm of twentieth-century science. By being "old-fashioned" and classical he was able to analyze the behavior of, in this instance, glaciers as no one had done before him. His work was one of the Harriman Expedition's primary contributions to science.

But in his monograph Gilbert did not stop with the three questions of climate, topography, and motion. In the second chapter he attempted to reconstruct the pre-Pleistocene horizon of Alaska and then the Pleistocene. He saw the pre-Pleistocene landscape as a vast peneplain lifted some six thousand feet high, then eroded into valleys and cliffs. Mountains arose, and then came the onset of the Ice Age—the Pleistocene. Glaciers formed and, like rivers seeking their base level (i.e., sea level), they cut down through existing valleys. If gravity, and hence velocity, spurred the glacier toward the sea at a rapid rate, it left slower-moving tributory glaciers high and "hanging." Gilbert never mentioned whether gravity, and hence velocity, had anything to do with the rate of the crustal uplift that went on simultaneously with the formation and movement of the glacier. In fact, he said very little about earthquakes and crustal displacement in relation to glaciers—a curious omission considering his familiarity with Powell's theories of Grand Canyon geology.

A final aspect of Gilbert's work was ingenious. At the suggestion of G. F. Becker he conducted a series of laboratory experiments to determine the extent and nature of glacial contact with the ocean floor below sea level. In these experiments he proved that glaciers do not displace their weight in sea water, but rather rest on a thin film or cushion of water and hence continue to operate upon the ocean floor.

Gilbert's work on the Harriman Expedition was a major contribution to glacial geology. He had described the Ice Age horizons and he had outlined the physical mechanics of glaciers and glacial action. Though he left some questions unanswered, his work was far ahead of its time. The expedition provided the occasion, but the man produced the result. This result in turn was due to Gilbert's very special binocular ability carefully to observe landforms in the concrete and at the same time to think of them in the abstract. He was unquestionably the most brilliant "scientific" aboard the *George W. Elder*, and his study of Alaskan glaciers represents the pinnacle of the expedition's contribution to science. Interestingly enough, Gilbert's work was really the exemplification of a team effort: he needed Harriman's boat and library; Gannett's maps; Muir's experience; and the photographs of Curtis, Merriam, and others. Only then could he "see what he had eyes to see."

There can be little question that Harriman's expedition was a serious scientific venture. In fact, in many ways it was an important milestone in the history of American science, not only because of the knowledge it produced, but also because it was the last of the grand oceanic exploring cruises that began in the days of Captain Cook and opened a new age of geographical discovery. There would, of course, be other voyages, dashes to the Poles, the discovery of a Northwest Passage by Amundsen, even cruises by dirigibles carrying companies of scientists, but they really belong to a new and different age. When the *Elder* docked at Seattle Harbor on July 30, 1899, the long Second Great Age of Discovery had nearly come to an end. The future lay not with the concrete description and savoring of nature's marvels, but with abstraction, analysis, non-common-sense logic, discrete systems, and multiple realities. They, too, would bring high adventure and moments of awe, but of a different kind in a different world.

APPENDIX*

MEMBERS OF THE HARRIMAN ALASKA EXPEDITION.

CLASSIFIED SUMMARY

Harriman family and servants 14
Scientific party 25
Artists 3
Photographers 2
Stenographers 2
Surgeon and assistant 2
Trained nurse 1
Chaplain 1
Hunters, packers, and camp hands 11
Officers and crew 65
                                                                              ———
                                                                              126

THE HARRIMAN FAMILY
Edward H. Harriman,
   patron of the expedition, Arden, N. Y.
Mrs. E. H. Harriman.
Miss Mary Harriman.

*From the Harriman Expedition *Reports*, Vol. I.

Miss Cornelia Harriman.

Carol Harriman.

Averell Harriman.

Roland Harriman.

W. H. Averell, Rochester, New York.

Mrs. W. H. Averell.

Miss Elizabeth Averell.

Miss Dorothea Draper, New York City.

THE SCIENTIFIC PARTY

Prof. William H. Brewer, Sheffield Scientific School, Yale University, New Haven, Conn.

John Burroughs, Ornithologist and Author, West Park, N. Y.

Wesley R. Coe, Ph.D., Assistant Professor of Comparative Anatomy, Yale University, New Haven, Conn.

Frederick V. Coville, Curator of the National Herbarium and Botanist of the U. S. Department of Agriculture, Washington, D. C.

Dr. William H. Dall, Paleontologist of the U. S. Geological Survey, and Honorary Curator of Mollusks, U. S. National Museum, Washington, D. C.

W. B. Devereux, Mining Engineer, Glenwood Springs, Colo.

Daniel G. Elliot, Curator of Zoology, Field Columbian Museum, Chicago, Ill.

Prof. Benjamin K. Emerson, Professor of Geology, Amherst College, Amherst, Mass.

Prof. B. E. Fernow, Dean of the School of Forestry, Cornell University, Ithaca, N. Y.

Dr. A. K. Fisher, Ornithologist, Biological Survey, U. S. Department of Agriculture, Washington, D. C.

Henry Gannett, Chief Geographer, U.S. Geological Survey, Washington, D. C.

G. K. Gilbert, Geologist, U. S. Geological Survey, Washington, D. C.

Dr. George Bird Grinnell, Editor, Forest and Stream, New York City.

Thomas H. Kearney, Jr., Assistant Botanist, U. S. Department of Agriculture, Washington, D. C.

Charles A. Keeler, Director of the Museum of the California Academy of Sciences, San Francisco, Calif.

Prof. Trevor Kincaid, Professor of Zoology, University of Washington, Seattle, State of Washington.

Dr. C. Hart Merriam, Chief of the Biological Survey, U.S. Department of Agriculture, Washington, D. C.

John Muir, Author and Student of Glaciers, Martinez, Calif.

Dr. Charles Palache, Mineralogist, Harvard University, Cambridge, Mass.

Robert Ridgway, Curator of Birds, U. S. National Museum, Washington, D. C.

Prof. William E. Ritter, President of the California Academy of Sciences and Professor of Zoology in the University of California, Berkeley, Calif.

De Alton Saunders, Botanist, South Dakota Experiment Station, Brookings, South Dakota.

Dr. William Trelease, Director of the Missouri Botanical Garden, St. Louis, Mo.

ARTISTS
R. Swain Gifford, New York City.
Frederick S. Dellenbaugh, New York City.

BIRD ARTIST
Louis Agassiz Fuertes, Ithaca, N. Y.

PHYSICIANS
Dr. Lewis Rutherford Morris, New York City.
Dr. Edward L. Trudeau, Jr., Saranac Lake, N. Y.

TAXIDERMISTS AND PREPARATORS
Leon J. Cole, Ann Arbor, Michigan.
Edwin C. Starks, Biological Survey, Washington, D. C.

PHOTOGRAPHERS
Edward S. Curtis, Seattle, Wash.
D. J. Inverarity, Seattle, Wash.

CHAPLAIN
Dr. George F. Nelson, New York City.

STENOGRAPHERS

Louis F. Timmerman, New York City.

Julian L. Johns, Washington, D. C.

SHIP'S OFFICERS

Captain, Peter Doran.

First Officer, Charles McCarty.

Pilot, J. F. Jordan.

Chief Engineer, J. A. Scandrett.

Steward, Joseph V. Knights.

At Orca, Prince William Sound, Capt. Omar J. Humphrey of the Pacific Steam Whaling Company joined the ship and accompanied the party to Bering Strait and back to Unalaska. His detailed knowledge of the coast proved of much value in navigating the ship. Mr. M. L. Washburn of the Alaska Commercial Company also joined the Expedition at Orca and went with it to Kadiak. On the return voyage Mr. J. Stanley-Brown of the North American Commercial Company came aboard at Dutch Harbor, Unalaska, and accompanied the party on the homeward voyage and the overland journey.

COMMITTEES

On the westward journey across the continent an organization was perfected, and the various activities of the Expedition were assigned to special committees, as follows:

EXECUTIVE COMMITTEE.

E. H. Harriman, *Chairman*	Bernhard E. Fernow
C. Hart Merriam, *Secretary*	Henry Gannett
Frederick V. Coville	G. K. Gilbert
Edward S. Curtis	George Bird Grinnell
Wm. H. Dall	Lewis R. Morris
W. B. Devereux	John Muir

COMMITTEE ON ROUTE AND PLANS.

E. H. Harriman, *Chairman*	W. B. Devereux
C. Hart Merriam, *Secretary*	Captain Peter Doran
Frederick V. Coville	Henry Gannett
Wm. H. Dall	G. K. Gilbert

Lewis R. Morris

COMMITTEE ON ZOOLOGY.

Wm. H. Dall, *Chairman*   C. Hart Merriam
Daniel G. Elliot   Robert Ridgway
          Wm. E. Ritter

COMMITTEE ON BOTANY.

F. V. Coville, *Chairman*   Wm. H. Brewer
          Wm. Trelease

COMMITTEE ON GEOLOGY.

G. K. Gilbert, *Chairman*   B. K. Emerson
          John Muir

COMMITTEE ON MINING.

W. B. Devereux, *Chairman*   Charles Palache

COMMITTEE ON GEOGRAPHY AND GEOGRAPHIC NAMES.

Henry Gannett, *Chairman*   B. K. Emerson
Wm. H. Dall   G. K. Gilbert
          John Muir

COMMITTEE ON BIG GAME.

Lewis R. Morris, *Chairman*   George Bird Grinnell
Daniel G. Elliot   Miss Mary Harriman

COMMITTEE ON LECTURES.

Henry Gannett, *Chairman*   Frederick V. Coville
Wm. H. Brewer   G. K. Gilbert
          George F. Nelson

COMMITTEE ON LIBRARY.

Mrs. Averell, *Chairman*   Miss Draper
Miss Harriman, *Secretary*   Miss Cornelia Harriman
Frederick V. Coville   E. L. Trudeau

COMMITTEE ON LITERATURE AND ART.

Mrs. Harriman, *Chairman*   Fred S. Dellenbaugh
G. F. Nelson, *Secretary*   Louis A. Fuertes
John Burroughs, *Historian*   R. Swain Gifford
Miss Averell   Miss Cornelia Harriman
          John Muir

COMMITTEE ON MUSIC AND ENTERTAINMENT.

B. E. Fernow, *Chairman* Fred S. Dellenbaugh

Miss Draper, *Secretary* Louis A. Fuertes

R. Swain Gifford

A NOTE ON THE SOURCES

Documenting the Harriman Expedition to Alaska presents a fascinating research problem: Manuscript materials for two of the primary figures in our story, E. H. Harriman and Edward S. Curtis, are extremely scarce for the period with which we have been concerned. Edward Harriman's personal papers were destroyed in a warehouse fire in 1913, and Edward S. Curtis kept no journals until 1927. Curtis was by nature an unselfconscious man, and certainly so during the 1890s, when he was just getting started as a photographer. He was much too busy making a living with his camera to keep diaries or reflect upon himself intensely. When he did make public statements about himself and his work, especially in reference to his "tricks of the trade," he was pointedly vague. Then, too, upon the breakup of his marriage and the dissolution of his studio in 1920, whatever records and correspondence he might have had in the studio passed to his estranged wife and then into oblivion. The whereabouts of these studio files is, of course, one of the great and tantalizing mysteries that confront all Curtis scholars. Another fascinating clue about Curtis' impressions may lie in the papers of his extraordinary assistants, D. J. Inverarity and especially Adolph Muhr, but we have been unable to locate or consult these. Finally, future Curtis scholars might be well advised to consult the corre-‘ spondence files of Edward's sibling Asahel, which can be found at the Washington State Historical Society and the University of Washington Northwest Collection.

But since our book is more than simply a study of Curtis, or even of Harriman, intriguing though those may be as possibilities, our research has led us down numerous other trails and into widely varying kinds of materials, including photographic archives, manuscript collections, obscure and long-forgotten periodicals, scientific treatises, and interviews not only with surviving relatives, but also with Averell Harriman, a survivor of the expedition itself. Thus our research, like our book, has been truly interdisciplinary. The Notes mark our many trails; what follows is a summary of our journey to date, which may be of interest as a starting point for future scholars.

I.   Manuscript and Archival Sources

1. The Seattle Public Library, Seattle, Washington
   Here may be found the most important documents on Curtis' early career, the Curtis letters to Harriet Leitch, 1948–1951. These were written at the end of Curtis' life in response to questions by Ms. Leitch, and take the form of reminiscences.

2. The University of Washington Special Collections, Seattle, Washington
   This contains an archive of Curtis photographs, those of other Alaskan photographers of the period, obscure and ephemeral journals, the Meany Papers, Trevor Kincaid's unpublished "Autobiography," and miscellaneous Curtisiana.

3. The Washington State Historical Society, Tacoma, Washington.
   The main repository for the Asahel Curtis photo archive.

4. The Lois Flury Collection, Port Townsend, Washington
   Primarily a collection of Curtis North American Indian photographs, but also contains some Asahel Curtis photos and some of Edward Curtis' earliest Indian photos from the 1890s.

5. The Asahel Curtis, Jr., Collection, Seattle, Washington
   Family photographs, Asahel Curtis photographs including his color work, and miscellaneous documents.

6. Florence Curtis Graybill Collection, Los Angeles, California
   Edward Curtis photographs, a typed list of his early Mount Rainier photographs of the 1890s, and a typed reminiscence of his extensive experience on Mount Rainier.

7. The Averell Harriman Collection, New York, New York
   A large souvenir scrapbook of the expedition, "Chronicles and Souvenirs," containing photographs, watercolors, maps, letters and poems composed by expedition members. The Harriman

collection also includes a manuscript version of John Burroughs' journal of the expedition.

8. The Southwest Museum, Los Angeles, California
   A copy of the *Souvenir Album*, the extensive and important correspondence between Curtis and Frederick Webb Hodge, and George Bird Grinnell's diary of the trip.

9. Yale University, New Haven, Connecticut
   The important Frederick Dellenbaugh diary of the expedition and William Brewer's pocket field journal and general journal of the expedition.

10. The Smithsonian Institution, Washington, D.C.
    William Healey Dall's journal and letters; also Thomas Kearney's unpublished "Reminiscences of the Harriman Expedition," in the Schmitt Collection.

11. The Library of Congress, Washington, D.C.
    The important C. Hart Merriam "home journals," "field journals," and Correspondence; the Albert Kendrick Fisher Letters.

12. The Huntington Library, San Marino, California
    The John Burroughs journal of the expedition.

13. The Amherst College Archives, Amherst, Massachusetts
    The B. K. Emerson Papers.

14. The Bancroft Library, Berkeley, California
    The Charles Keeler Papers and the C. Hart Merriam Letterbooks. The Bancroft Library also has the Charles and Louise Keeler drawings from the Harriman Expedition, Charles Keeler's photographs, and C. Hart Merriam's assemblage of most of the photos from the expedition, including fifty-one hitherto unknown E. S. Curtis photographs.

15. The *Souvenir Album*
    The Harriman Expedition *Souvenir Album* may be found at: University of Texas, Austin; U.C.L.A.; Yale; The Smithsonian Institution; The University of Washington Special Collections; the Bancroft Library; and the Library of Congress.

16. The Holt-Atherton Center for Western Studies, University of the Pacific, Stockton, California. The John Muir Papers and Notebooks. This collection also includes unknown E. S. Curtis photographs and a striking set by Grove Karl Gilbert.

II. Printed Primary Sources

1. C. Hart Merriam, ed., *Harriman Alaska Expedition*, 13 vols.
   (New York: Doubleday, Page and Company, and Washington, D.C.: Smithsonian Institution, 1901–1914).
2. Letters, diaries, and reminiscences
   John Muir, *John of the Mountains: Unpublished Journals of John Muir*, Linnie Marsh Wolfe, ed. (Boston: Houghton Mifflin Company, 1938).
   Clara Barrus, ed., *The Heart of Burroughs' Journals* (Boston and N.Y.: Houghton Mifflin Company, 1938).
   John Burroughs, *My Boyhood* (Garden City, N.Y.: Doubleday, Page and Company, 1928).
3. Contemporary Books and Articles
   John Muir, *Travels in Alaska* (Boston: Houghton Mifflin Company, 1915)
   ———, *Edward Henry Harriman* (New York: Doubleday, Page and Company, 1912)
   Cyrus C. Adams, "The World's Great Railroad Enterprises," *World's Work* 13:8423 (January 1907).
   William Healey Dall, "Alaskan Notes," *The Nation* 69:127 (August 17, 1899).
   ———, "Alaska's Railroads," *Review of Reviews* 24:346 (March 1904).
   ———, "Discoveries in Our Arctic Region," *World's Work* I:149 (December 1900).
   ———, "Alaska Revisited," *The Nation* 6:6–7 (July 4, 1895).
   Edward S. Curtis, "The Amateur Photographer," in *The Western Trail* vol. I, no. 3 (January 1900): 379–80; vol. 1, no. 4 (February 1900):272–74; vol. I, no. 5 (March–April 1900):379–80; vol. 1, no. 6 (May 1900):468–69.
   ———, "The Rush to the Klondike over the Mountain Passes," *The Century Magazine* vol. LV, no. 5 (March 1898):692–97.
   The Argus (Seattle), "Will Go to the Klondike," December 18, 1897; "Curtis' Photographic Studio," December 17, 1898; "Curtis and Guptil," December 19, 1846.
   A. F. Muhr, "A Gum-Bichromate Process for Obtaining Colored Prints from a Single Negative," *Camera Craft: A Photographic Monthly* (San Francisco) vol. XIII, no. 2 (August 1906):277 ff. Also in same issue, "Mr. Curtis and His Able

Co-worker, Mr. Muhr," pp. 308–309.

C. M. Keyes "Harriman" *World's Work* vol. 13, four-part series, January–April 1907.

Jeremiah Lynch, *Three Years on the Klondike*, (London: Edward Arnold, 1904).

C. L. Marlatt, "Review of Harriman *Reports*, Vols. VIII and IX," *Science*, vol. XX, no. 514 (November 4, 1903):601–603.

W. H. Dall, "Review of Harriman *Reports*, Vol. X," *Science*, vol. XX, no. 510 (October 7, 1904):462–64.

W. J. M., "Review of Harriman *Reports*, Vols I, II," *Science*, vol. XIV, no. 360 (November 22, 1901):807–808.

Israel C. Russell, "Review of Harriman *Reports*, Vols. III, IV," *Science*, vol. XIX, no. 490 (May 20, 1904):917–9.

Lucien M. Underwood, "Review of Harriman *Reports*, Vol. V," *Science*, vol. XIX, no. 490 (June 17, 1904):917–19.

C. A. Koford, "Our Land of the Midnight Sun" (Review of Vols. I and II of Harriman *Reports*), *The Dial*, vol. 31, no. 273 (October 16, 1901):273–76.

Henry Gannett, "The Harriman Alaska Expedition," *National Geographic*, vol. 10, No. 507, pp. 507–12.

4. Newspapers
   Portland *Oregonian*, May 1899
   Boise *Statesman*, May 1899
   New York *Herald*, April 1899
   New York *Daily Tribune*, April–August 1899
   Seattle *Times*, 1899
   Seattle *Intelligencer*, 1899

5. Photographic Collections
   Smithsonian Institution, Washington, D.C.
   University of Texas, Austin, Texas
   Washington State Historical Society
   University of Washington Northwest and Photographic Collections, Seattle, Washington
   Lois Flury Collection, Port Townsend, Washington
   Asahel Curtis, Jr., Collection, Seattle, Washington
   Averell Harriman Collection, New York, New York
   The Southwest Museum, Los Angeles, California
   The Bancroft Library, Berkeley, California
   Joslyn Museum, Omaha, Nebraska

III. Authorities

The number of books and articles on the history of photography
and on Alaska is substantial and growing every day. The following
represent some of the studies we found most useful: George Kennan,
E. H. Harriman: A Biography, 2 vols. (Boston: Houghton Mifflin
Company, 1922); George Kennan, *E. H. Harriman's Far Eastern Plans*
(New York: The Country Life Press, 1917); Robert A Lovett, *Forty
Years After: An Appreciation of the Genius of Edward Henry Harriman*
(Princeton: Princeton University Press, 1949); Keir B. Sterling, *Last
of the Great Naturalists: The Career of C. Hart Merriam* (New York:
Arno Press, 1977); Hall Young, *Alaska Days with John Muir* (New
York: Fleming H. Revell Company, 1915); Morgan B. Sherwood,
The Exploration of Alaska (New Haven & London: Yale University
Press, 1965); Gustavus Myers, *History of the Great American Fortunes*,
vol. III (Chicago: Charles Kerr & Company, 1910); John F. Reiger,
ed., *The Passing of the Great West* (New York: Winchester Press, 1972);
Pierre Breton, *Klondike: The Last Great Gold Rush* (Toronto: McClel-
land and Stewart, Ltd., 1972); Vijhalmur Stefansson, *Northwest to
Fortune* (New York: Duell, Sloan & Pearce, 1958); Ted C. Hinck-
ley, *The Americanization of Alaska, 1867–1897* (Palo Alto, California:
Pacific Books, 1972); William R. Hunt, *Alaska: A Bicentennial History*
(New York: W. W. Norton, 1976); William R. Hunt, *North of 53*
(New York: MacMillan Company, 1973); William R. Hunt, *Arctic
Passage* (New York: Charles Scribner's Sons, 1975), Stephen Pyne,
"Grove Karl Gilbert: A Biography of American Geology," unpub-
lished Ph.D. dissertation, University of Texas American Civilization
Program; W. F. Bade, *The Life and Letters of John Muir*, 2 vols. (Bos-
ton: Houghton Mifflin Company, 1929); Linnie Marsh Wolfe, *Son of
the Wilderness: The Life of John Muir* (Madison: University of Wiscon-
sin Press, 1980).

On Curtis and related photographers: Florence Curtis Graybill and
Victor Boesen, *Edward Sheriff Curtis, Visions of a Vanishing Race* (New
York: Thomas Y. Crowell, 1976); Victor Boeson and Florence Cur-
tis Graybill, *Edward Sheriff Curtis, Photographer of the North American
Indians* (New York: Dodd, Mead & Company, 1977); Bill Holm and
George Irving Quimby, *A Pioneer Cinematographer in the Pacific North-
west: Edward S. Curtis in the Land of the War Canoes* (Seattle and Lon-
don: University of Washington Press, 1980); Alvin Josephy, "The
Splendid Indians of E. S. Curtis," *American Heritage* vol. 25, no. 2

(February 1974):40–59, 96–97; Mick Gidley, *The Vanishing Race* (New York: Taplinger Publishing Company, 1977); Alton A. Lindsey, "The Harriman Alaska Expedition of 1899, Including the Identities of Those in the Staff Picture," *BioScience* vol. 28, no. 6 (1978):383–86; Mick Gidley, "Edward S. Curtis Speaks . . ." *History of Photography, an International Quarterly* vol. 2, no. 4 (October 1978):347–54, Ralph W. Andrews, *Curtis' Western Indians* (Seattle: Superior Publishing Company, 1962); Ralph W. Andrews, *Photographers of the Frontier West, 1875 to 1910—Their Lives and 'Their Works* (Seattle: Superior Publishing Company, 1965), pp. 38–59; Clark Worswick and Jonathan Spence, *Imperial China: Photographs, 1850–1912* (New York: Crown, 1978); Clark Worswick, *Japan: Photographs, 1854–1905* (New York: Knopf, 1979); A. D. Coleman and T. C. McLuhan, *Portraits from North American Indian Life* (New York: Promontory Press, 1972); Edmond S. Meany, "Hunting Indians with a Camera," *World's Work* 15 (March 1912): 10004–11; Weston J. Naef and James Wood, *Era of Exploration: The Rise of Landscape Photography in the American West, 1860–1885* (Boston: New York Graphic Society, 1975); William H. Goetzmann, *Exploration and Empire: The Explorer and the Scientist in the Winning of the American West* (New York: Knopf, 1966); William Welling, *Collector's Guide to Nineteenth-Century Photographs* (New York: Collier Books, 1976), Karen and William Current, *Photography and the Old West* (New York and Fort Worth Texas: Harry N. Abrams/Amon Carter Museum of Western Art, 1978); Beaumont Newhall, *The History of Photography* (Boston: New York Graphic Society, 1978); John Wilmerding et al., *American Light: The Luminist Movement, 1850–1875, Paintings, Drawings, Photographs* (Washington, D.C.: The National Gallery of Art, 1980); William H. Goetzmann, "Paradigm Lost" in Nathan Reingold, ed., *The Science in the American Context: New Perspectives* (Washington, D.C.: Smithsonian Institution Press, 1979); Nicolette Ann Bromberg, "Clarence Leroy Andrews and Alaska . . . One of Alaska's Foremost Historical Writers and Photographers," *The Alaska Journal* vol. 6, no. 2 (Spring 1976): 66–77; John Starrets, "Asahel Curtis: The Forgotten Brother," *Images, Seattle Sun Photo Supplement, 1979*; David Sucher, ed., Murray Morgan, foreword, Wes Uhlman, afterword, *The Asahel Curtis Sampler: Photographs of Puget Sound Past* (Seattle: Puget Sound Access, 1973); Richard Frederick, "Photographer Asahel Curtis, Chronicler of the Northwest," *American West* (December 1980):26–40.

NOTES

Preface and Acknowledgments

xiv "a capital ship": Charles E. Carryl, "The Walloping Window-blind," in Carolyn Wells, ed., *A Nonsense Anthology* (New York: Dover Publications, 1958), p. 123.

Chapter One
The Dynamics of Philanthropy

5 chairman of the board: For accounts of Edward Harriman's life and career, *see* George Kennan, *Biography of E. H. Harriman*, 2 vols. (Boston: Houghton Mifflin Company, 1922); C. M. Keyes, "Harriman," a series of four articles in *World's Work* vol. 13: January 1907, pp. 8455–64; February 1907, pp. 8537–52; March 1907, pp. 8651–64; and April 1907, pp. 8791–8803.

5 an extended vacation: Keir B. Sterling, *The Last of the Naturalists: The Career of C. Hart Merriam* (New York: Arno Press, 1977), p. 119.

5 strength of the creature: John Burroughs, John Muir, and George Bird Grinnell, *Harriman Alaska Series* vol. I, ed. C. Hart Merriam (New York: Doubleday, Page and Company, 1902), p. xxi.

6 "nose for money": Robert Abercrombie Lovett, *Forty Years After: An Appreciation of the Genius of Edward Henry Harriman* (Princeton: Princeton University Press, 1949), p. 12.

6 "he was looked at askance": Otto Kahn, *Our Economic and Other Problems* (New York: George H. Doran Company, 1920), p. 19.

6 Pacific Coast Steamship Company: *See* Septima M. Collis, *A Woman's Trip to Alaska* (New York: Cassell Publishing Company, 1890); *Four Thousand Miles North and South From San Francisco* (San Francisco: Pacific Coast Steamship Company, 1896).

8 circle the world: *See* George Kennan, *E. H. Harriman's Far Eastern Plans* (New York: Country Life Press, 1917); Vilhjalmur Stefansson, *Northwest to Fortune* (New York: Duell, Sloan and Pearce, 1958); Lovett, *Forty Years After* (1949).

8 all at his expense: C. Hart Merriam, "Home Journals," March–May 1899, Smithsonian Institution Archives. *See also* Sterling (1977), pp. 119–21.

9 the expedition's party: *Cosmos Club Membership Book* (Washington, D.C.: Cosmos Club, 1917).

10 Rocky Mountain West: The information that follows is derived primarily from *The Dictionary of American Biography* (New York: Charles Scribner's Sons, 1928–present) and *Biographical Memoirs*, 52 vols. (Washington, D.C.: National Academy of Science, 1928–1961).

10 "little love for the North": "Robert Ridgway," *Biographical Memoirs* vol. 15 (1932), p. 57.

10 nature of the expedition: John Muir, *Edward Henry Harriman* (New York: Doubleday, Page and Company, 1912), pp. 8–10.

11 his friend Walt Whitman: *See* Paul Brooks, "The Two Johns—Burroughs and Muir," *Sierra* (September/October 1980), p. 53.

11 among his favorites: Edmund Morris, *The Rise of Theodore Roosevelt* (New York: Coward, McCann and Geoghegan, Inc., 1979), p. 300.

11 "your college authorities": C. Hart Merriam to B. K. Emerson, April 14, 1899. B. K. Emerson letters, Amherst College Archives.

12 remote Arctic glaciers: Charles Palache to B. K. Emerson, May 2, 1899. B. K. Emerson letters.

12 photographer for the voyage: *See* Chapter 14 for an account of the meeting between Edward Curtis and C. Hart Merriam, George Bird Grinnell, and Gifford Pinchot in 1897.

12 at The Cooper Union: "Robert Swain Gifford, 1840–1905," catalogue accompanying exhibition of Gifford's paintings, New Bedford Whaling Museum, New Bedford, 1974.

12 to Salt Lake City: "The Exploration of the Colorado River and

the High Plateaus of Utah in 1871–72," *Utah Historical Quarterly* XVII (1949), pp. 497–99.

13 "we can do a great deal": Robert Swain Gifford to Frederick S. Dellenbaugh, April 29, 1899. Dellenbaugh papers, Yale Western Americana Collection.

13 "I want to impress upon you": Mary Boynton Fuertes, *Louis Agassiz Fuertes* (New York: Oxford University Press, 1956), p. 43.

13 "with very full portfolios": *Ibid.*, pp. 43–44.
"invasion of Alaska": "Many Scientists to Invade Alaska," New York *Herald*, April 23, 1899.

14 date to the press: Robert Swain Gifford to Frederick S. Dellenbaugh, May 11, 1899. Dellenbaugh papers.

Chapter Two
Across the Continent in Pullman's Palace Cars

16 Harriman connection on time: William Brewer Diary, "Harriman Alaska Expedition, 1899," 2 vols., Yale University Library. The narrative of the Harriman Expedition is based primarily on the journals of William Brewer, Frederick S. Dellenbaugh ("Harriman Alaska Expedition," Yale Western Americana Collection), John Burroughs ("Album of the Alaska Expedition," Governor Averell Harriman family estate), John Burroughs et al, *Harriman Alaska Series* vol. I, ed. C. Hart Merriam (New York: Doubleday, Page and Company, 1902), and C. Hart Merriam ("Harriman Alaska Expedition," Library of Congress).

17 the railroad king: Dellenbaugh Diary.

18 upon meeting him: Thomas H. Kearney, "Reminiscences of the Harriman Expedition," unpublished manuscript, Waldo Schmitt Papers, Smithsonian Institution Archives.

18 "Have I made a mistake": Clara Barrus, ed., *The Heart of Burroughs' Journals* (Boston and New York: Houghton Mifflin Company, 1928), p. 212.

19 "perfected its plans": Merriam Diary.

19 "in places the country": Burroughs, *Harriman Alaska Series* vol. I, p. 6.

21 "Uncle John.": Boynton, *Louis Agassiz Fuertes*, p. 53.

21 "like a nightmare.": Burroughs, *Harriman Alaska Series* vol. I, p. 3.

24 "the man of the hour": "Distinguished Party Will Visit Boise Today," Boise *Statesman*, May 28, 1899. Clipping held in Dellenbaugh Diary.

25 "signs mean nothing": Dellenbaugh Diary.

26 "withdrawn into her bower": Burroughs, *Harriman Alaska Series* vol. I, p. 16.

26 "charming companion": John Muir, *John of the Mountains: Unpublished Journals of John Muir*, ed. Linnie Marsh Wolfe (Boston: Houghton Mifflin Company, 1938), p. 379.

26 wife and young daughter: Charles Keeler to Louise Keeler, May 28, 1899. Keeler letters, Bancroft Library.

26 "I start tomorrow": John Muir to Walter Hines Page, May 1899, quoted in W. F. Bade, *The Life and Letters of John Muir* vol. II (Boston: Houghton Mifflin Company, 1924), pp. 320–21.

27 "people look at what": Muir, *John of the Mountains*, p. 379.

27 "is almost as great": Charles Keeler to Louise Keeler, May 27, 1899. Keeler letters.

28 Muir's favorite subject: Muir, *John of the Mountains*, pp. 379–80.

28 provisions for the expedition: Charles Keeler to Louise Keeler, May 29, 1899. Keeler letters.

28 from his exuberant friend: Muir, *John of the Mountains*, p. 380.

29 "a striking illustration": Editorial, Portland *Oregonian*, May 30, 1899. Clipping held in Dellenbaugh Diary.

29 simple and unassuming: John Muir, *Edward Henry Harriman* (New York: Doubleday, Page and Company, 1912), p. 10. Also Charles Keeler to Louise Keeler, June 4, 1899. Keeler letters.

30 Muir modestly disclaimed: Muir, *John of the Mountains*, p. 380.

30 rain-swept wharf: A graphophone was a machine that both recorded and reproduced sound.

Chapter Three
"The Best Object Lessons to Be Found on the Coast"

31 "Of course we are a trifle": Charles Keeler to Louise Keeler, May 31, 1899. Keeler letters.

32 the expedition's schedule and routes: John Muir to family, June

1, 1899, quoted in Bade, *The Life and Letters of John Muir* vol. II, p. 322.

33 on board the ship: George Bird Grinnell Diary, Harriman Expedition to Alaska, May–August 1899, Southwest Museum, Los Angeles, California.

33 submerged the sharp rocks: Brewer Diary.

36 a letter to his wife: Charles Keeler to Louise Keeler, June 3, 1899. Keeler letters.

37 alone in the wilderness: Boynton, *Louise Agassiz Fuertes*, p. 43.

37 his companions on board: William Healey Dall Diary, May–August 1899, Smithsonian Institution Archives.

38 from Fisher: Boynton, *Louis Agassiz Fuertes*, p. 45.

38 "civilizing the savages": Burroughs, "Album of the Harriman Alaska Expedition," Harriman family estate.

40 "ladies and gentlemen": Charles Keeler to Louise Keeler, June 4, 1899. Keeler letters.

40 "really the father": Burroughs, *Harriman Alaska Series* vol. I, p. 26.

40 political rights by 1915: Ivan Doig, "The Tribe That Learned the Gospel of Capitalism," *The American West* vol. XI (March 1974), pp. 42–47.

41 "one of the best object lessons": Burroughs, *Harriman Alaska Series* vol. I, pp. 24–26.

41 "if they could be taught": "Mr. Harriman Talks of Alaska," New York *Daily Tribune*, August 14, 1899, p. 12.

41 "It took many years": Burroughs, *Harriman Alaska Series* vol. I, p. 154.

43 rate of physical exertion: Albert Kendrick Fisher to Walter Fisher, June 6, 1899. Albert Kendrick Fisher letters, Library of Congress.

44 twilight "alpenglow": Brewer Diary.

44 "you ought to have been": Dellenbaugh Diary.

44 young scientist's face: Charles Keeler to Louise Keeler, June 4, 1899. Keeler letters.

44 "always trying": Charles Keeler to Louise Keeler, June 4, 1899. Keeler letters.

45 "fearfully and wonderfully": John Burroughs to Julian Burroughs, in John Burroughs, *My Boyhood* (New York: Doubleday, Page and Company, 1922), p. 221.

45 "Seaweed Saunders": John Muir to "The Big Four," August

30, 1899, quoted in Bade, *The Life and Letters of John Muir* vol. II, p. 332.

45 vineyards at home: Burroughs, *My Boyhood*, p. 219.

46 "peacefully decaying": William Healey Dall, "Alaskan Notes," *The Nation* 69:128 (August 17, 1899).

46 "dirty, miserable town": Charles Keeler to Louise Keeler, June 5, 1899. Keeler letters.

47 "very democratic": John Burroughs to Julian Burroughs, June 5, 1899, quoted in Burroughs, *My Boyhood*, p. 220.

48 "turning over the leaves": John Muir to his wife and daughters, June 6, 1899, quoted in Bade, *The Life and Letters of John Muir* vol. II, p. 323.

48 of the Patterson: Dellenbaugh Diary.

48 "the smallest lamb": Burroughs, "Album of the Harriman Alaska Expedition," Harriman family estate.

50 back to rescue him: Grinnell Diary.

50 "Say, you're quite": Dellenbaugh Diary.

51 periods of sunlight: Edward S. Curtis to Harriet Leitch, November 17, 1950. Curtis-Leitch letters

51 "houses of the dead": Merriam Diary.

52 writing found fans: John Burroughs to Julian Burroughs, June 17, 1899, quoted in Burroughs, *My Boyhood*, p. 223.

53 "a soft hum": Burroughs, "Album of the Harriman Alaska Expedition."

53 "As long as this dog": Trevor Kincaid, "Unpublished Autobiography," manuscript, University of Washington Archives.

Chapter Four
The Stuff of Legends:
Skagway, the Gold Rush, and Dead Horse Trail

56 "a nest of ants": S. Hall Young, *Alaska Days with John Muir* (New York: Fleming H. Revell Company, 1915), p. 210.

56 the White Pass managers: Dellenbaugh Diary.

56 rash and foolish: "Alaska's Railroads," *Review of Reviews* 29:346 (March 1904).

57 mad crowds of miners: Young, *Alaska Days with John Muir*, p. 211.

58 "gazed intently at": Burroughs, "Album of the Harriman Alaska Expedition."

59 "I sat at his feet": Young, *Alaska Days with John Muir*, p. 13.
59 Young in 1879: *Ibid.*, pp. 95–122.
59 "instead of the music": *Ibid.*, pp. 216–17.
60 "one of the best illustrations": Dellenbaugh Diary.
60 "suburbs" of log cabins: Boynton, *Louis Agassiz Fuertes*, p. 45.
61 "ribs of the earth": Burroughs, *Harriman Alaska Series* vol. I, p. 32.
61 to the small band: Boynton, *Louis Agassiz Fuertes*, pp. 45–46.
62 all their specimens: George Bird Grinnell to Albert Kendrick Fisher, date unknown, Albert Kendrick Fisher letters.
63 "The terrible and the sublime": Burroughs, *Harriman Alaska Series* vol. I, p. 32.
64 "bent on getting": Dellenbaugh Diary.
65 "illustration of the needless": Grinnell Diary.
65 "How the devil": Dellenbaugh Diary.
65 glanced up at the train: Burroughs, *Harriman Alaska Series* vol. I, p. 34.
65 "town of the future": Dall, "Alaskan Notes," p. 128.
66 "what a show": John Muir, *Travels in Alaska* (Boston: Houghton Mifflin and Company, 1915), p. 293.
66 "when the Alaskan beauties": "The Harriman Expedition," New York *Daily Tribune*, August 15, 1899, p. 9.
66 "Alaska's grandeur": Henry Gannett, *Harriman Alaska Series* vol. II, ed. by C. Hart Merriam (New York: Doubleday, Page and Company, 1902), p. 277.
66 "enormous growth": "Mr. Harriman Talks of Alaska," New York *Daily Tribune*, August 14, 1899, p. 12.
66 "enormous quantities": Bernhard Fernow, *Harriman Alaska Series* vol. II, p. 254.

Chapter Five
John Muir's Country

69 of ice, he thought: Burroughs, *Harriman Alaska Series* vol. I, p. 36.
69 known as Endicott Valley: Merriam Journal and Burroughs, *Harriman Alaska Series* vol. I, pp. 38–40.
71 "The old glacier": Young, *Alaska Days with John Muir*, p. 116.
71 wolves never attacked: Muir, *Travels in Alaska*, p. 297.
73 "We saw the world-shaping": Burroughs, *Harriman Alaska Series* vol. I, p. 39.

73   his young company: Muir, *John of the Mountains*, p. 384.

74   back to the *Elder*: Merriam Diary.

76   "all the howling.": Burroughs, *Harriman Alaska Series* vol. I, p. 40.

80   day's hard work: Boynton, *Louis Agassiz Fuertes*, p. 47.

81   "solitude as of interstellar": Burroughs, *Harriman Alaska Series* vol. I, p. 47.

83   "a latent and terrible": Charles Keeler notes on Muir Glacier, MS, Keeler papers, Bancroft Library.

83   "glacier worms": Kincaid, "Unpublished Autobiography."

83   "It seemed": Boynton, *Louis Agassiz Fuertes*, p. 48.

85   "wild to get on": Muir, *John of the Mountains*, p. 386.

85   "longed to break away": Young, *Alaska Days with John Muir*, p. 220.

85   "Running the launch": C. Hart Merriam, quoted in Kennan, *Biography of E. H. Harriman* vol. I, p. 203.

85   "Who are we?": Mrs. Edward Harriman, "Chronicles and Souvenirs," album of the Harriman Expedition, Harriman family estate.

Chapter Six
"Cradled in Custom"

86   "Cradled in Custom": Robert Service, "The Call of the Wild," *The Spell of the Yukon* (Philadelphia: E. Stern and Company, 1907).

87   utterly "hideous": Charles Keeler to Louise Keeler, June 16, 1899. Keeler letters.

88   these Indians' houses: Dellenbaugh Diary.

88   he had ever found: Burroughs, *Harriman Alaska Series* vol. I, p. 50.

89   under Russian governance: William Healey Dall, "Alaska Revisited," *The Nation* 61: 6–7 (July 4, 1895).

90   hunting and fishing outings: Ted C. Hinckley, *The Americanization of Alaska, 1867–1897* (Palo Alto: Pacific Books, 1972), p. 229.

90   out of Hades: John Burroughs to Julian Burroughs, June 17, 1899, in Burroughs, *My Boyhood*, p. 222.

90   "murdered a mother deer": Muir, *John of the Mountains*, p. 388.

91   sacred tribal rituals: Grinnell Diary.

91   rare "hermaphrodite" blossom: Dellenbaugh Diary.

91 "society affair": John Muir to his wife and daughters, June 1899, quoted in Bade, *The Life and Letters of John Muir* vol. II, p. 324.

92 of the morning: Hinckley, *The Americanization of Alaska*, pp. 228–30.

92 "running a gauntlet": Dellenbaugh Diary.

92 "there is no snobbery": Charles Keeler to Louise Keeler, June 16, 1899. Keeler letters.

92 "sunny" Sitka: Brewer Diary.

93 vacationing on the ship: John Muir to his wife and daughters, June 1899, quoted in Bade, *The Life and Letters of John Muir* vol. II, p. 325.

93 his business problems: Charles Keeler to Louise Keeler, June 16, 1899. Keeler letters.

93 "I should be quite": John Burroughs to Julian Burroughs, June 17, 1899, in Burroughs, *My Boyhood*, p. 223.

Chapter Seven
Malaspina's Mistake: Yakutat Bay

95 "that it did not disturb": Burroughs, *Harriman Alaska Series* vol. I, p. 52.

96 "the most forbidding": Dellenbaugh Diary.

98 name of the bay: Muir, *John of the Mountains*, p. 390.

98 half a mile: John Muir to his wife, June 24, 1899, quoted in Bade, *The Life and Letters of John Muir* vol. II, p. 327.

99 "a special playground": Burroughs, *Harriman Alaska Series* vol. I, p. 56.

99 from the Alaskan natives: Mrs. Edward Harriman, "Chronicles and Souvenirs."

100 "one of the filthiest": Charles Keeler to Louise Keeler, June 24, 1899. Keeler letters.

100 of the carcasses: Grinnell Diary.

100 "watching and waiting": Burroughs, *Harriman Alaska Series* vol. I, p. 61.

100 "involve the imagination": John Burroughs to Julian Burroughs, November 30, 1899, in Burroughs, *My Boyhood*, p. 231.

100 "barking or half-howling": Muir, *John of the Mountains*, p. 393.

101 they had to offer: Merriam Diary.

102 he had wandered: Charles Keeler to Louise Keeler, June 21, 1899. Keeler letters.

102 "rejoiced to see": Merriam Diary.

102 "little rest from the oars.": Kearney, "Reminiscences of the Harriman Expedition."

103 improved his health: Charles Keeler to Louise Keeler, June 24, 1899. Keeler letters.

103 "like the spirit": Burroughs, *Harriman Alaska Series* vol. I, p. 64.

104 "upon the South": Grinnell Diary.

Chapter Eight
"The Map's Void Spaces"

105 "The Map's Void Spaces": Robert Service, "The Call of the Wild," *The Spell of the Yukon* (Philadelphia: E. Stern and Company, 1907).

106 "Men in this business": Muir, *John of the Mountains*, p. 394.

107 "They were from": Burroughs, *Harriman Alaska Series* vol. I, pp. 65–66.

107 would appall him: John Burroughs, Journal of Harriman Alaska Expedition, Huntington Library, Los Angeles, California.

107 "homely, slow": Burroughs, *Harriman Alaska Series* vol. I, p. 66.

107 in the Klondike: Dellenbaugh Diary.

109 "great ice chest": Burroughs, "Album of the Alaska Expedition."

110 "We shall discover": Muir, *John of the Mountains*, p. 396.

110 "every little fish pond": *Ibid.*

110 "Nothing in his way": Muir, *Edward Henry Harriman*, p. 13.

110 "at full speed . . .": Muir, *John of the Mountains*, p. 396.

111 into the wilderness: Burroughs, *Harriman Alaska Series* vol. I, p. 72.

112 pulling through safely: Brewer Diary.

113 "like jugglers": Burroughs, *Harriman Alaska Series* vol. I, p. 74.

113 out of the region: Clarence C. Hulley, *Alaska: Past and Present* (Portland: Binfords and Mort, 1970), p. 218. *See also* William R. Hunt, *Alaska* (New York: W. W. Norton and Company, 1976), pp. 79–81.

113 "By and by": quoted in Hinckley, *The Americanization of Alaska*, p. 194.

115 "no more ice time": Muir, *John of the Mountains*, p. 399.

115 "standing up": Merriam Diary.

Chapter Nine
"Dog Dirty and Loaded for Bear"

116 "Dog Dirty and Loaded for Bear": Robert Service, "The Shooting of Dan McGrew," *The Spell of the Yukon* (Philadelphia: E. Stern and Company, 1907).

116 hard rowing away: Burroughs, *Harriman Alaska Series* vol. I, p. 78.

117 "gun laden": Muir, *John of the Mountains*, p. 400.

117 enforce its laws: Grinnell Diary.

117 of the island: Burroughs, "Album of the Alaska Expedition."

118 the Harriman visitors: Dellenbaugh Diary.

118 killed the largest bear: Merriam Diary.

118 "No green mountains": Muir, *John of the Mountains*, pp. 400–401.

119 "sauntering as if": *Ibid.*, p. 400.

119 "pleasant and cordial": Dellenbaugh Diary.

119 "like beaver fur": Merriam Diary.

120 "a real big genuine": *Ibid.*

121 "lest the bear": Kincaid, "Unpublished Autobiography."

123 "the first to fight": Mrs. Edward Harriman, "Chronicles and Souvenirs."

123 "Fourth of July Ode": Dellenbaugh Diary.

124 "Decidedly out of place": *Ibid.*

124 "Our Banner": Mrs. Edward Harriman, "Chronicles and Souvenirs."

126 "mother and child." Muir, *John of the Mountains*, p. 402.

126 "had heard of": *Ibid.*

126 "Everything": Charles Keeler to Louise Keeler, July 3, 1899. Keeler letters.

128 round-the-world line: *See* Kennan, *E. H. Harriman's Far Eastern Plans.*

Chapter Ten
The Siberian Connection

129 speech from memory: Brewer Diary.

130 "felt more deeply": Merriam Diary.

130 "grim, unmanageable": Muir, *John of the Mountains*, p. 404.

131 "one man sent": Dellenbaugh Diary.

132 eggs for breakfast: Burroughs, *Harriman Alaska Series* vol. I, p. 82.

132 "Where are you": Charles Keeler, "Friends Bearing Torches," unpublished manuscript, quoted in Muir, *John of the Mountains*, p. 405.

132 "a veritable flower": Merriam Diary.

133 "bellowing and roaring": *Ibid.*

134 "handsome little fellows": Muir, *John of the Mountains*, p. 406.

134 "pay his respects": Dellenbaugh Diary.

135 "roared savagely": Merriam Diary.

135 "so as to render": Dellenbaugh Diary.

135 for future resources: Grinnell Diary.

136 "A great day": Merriam Diary.

136 "he was romping": Burroughs, *Harriman Alaska Series* vol. I, p. 99.

136 "pay for this voyage": Charles Keeler to Louise Keeler, July 18, 1899. Keeler letters.

137 "were not attractive": Merriam Diary.

137 "The contact": Muir, *John of the Mountains*, p. 408.

137 "the most barren": Merriam Diary.

137–40 "they sat down": Burroughs, *Harriman Alaska Series* vol. I, p. 101.

140 the "midnight sun": Dellenbaugh Diary.

141 "looked much cleaner": *Ibid.*

142 "a merry gypsy": Muir, *John of the Mountains*, p. 408.

144 "to see the demolishment": *Ibid.*, p. 409.

145 his fascinated listener: William Brewer Pocket Journal, June–August 1899, Yale University Library.

Chapter Eleven
"I Don't Give a Damn
If I Never See Any More Scenery"

147 "absolutely nothing": Dellenbaugh Diary.

147 strange-looking beasts: Merriam Diary.

148 "I've had good": Boynton, *Louis Agassiz Fuertes*, p. 51.

148 "perfectly ready": Albert Kendrick Fisher to Walter Fisher, June 29, 1899. Albert Kendrick Fisher papers.

148 "There is something": Brewer Diary.

148 "truly the most": Boynton, *Louis Agassiz Fuertes*, pp. 51–52.

149 easy to believe: Dellenbaugh Diary.

149 "Pot that": Boynton, *Louis Agassiz Fuertes*, p. 52.

149 "we had the finest": *Ibid.*

150 "snowy owl": Muir, *John of the Mountains*, p. 411.

151 "Your family": Dellenbaugh Diary.

151 "pyramid standing": Muir, *John of the Mountains*, p. 413.

152 "much twaddle": *Ibid.*

152 "Let nobody think": Dellenbaugh Diary.

152 "You are missing": Kearney, "Reminiscences of the Harriman Expedition."

153 "the scientists have a way": "Mr. Harriman Talks of Alaska," New York *Daily Tribune*, August 14, 1899, p. 12.

153 "A headless man": Mrs. Edward Harriman, "Chronicles and Souvenirs."

153 "so fine no drops": Brewer Diary.

154 "all the trout": Merriam Diary.

154 "was one of": Muir, *John of the Mountains*, p. 412.

154 "all cheered him": *Ibid.*

155 "they have not": Merriam Diary.

155 "getting so numerous": Dellenbaugh Diary.

156 "The fur seals": Grinnell Diary.

156 "partly," he wrote: Dellenbaugh Diary.

158 "were probably anxious": *Ibid.*

159 "returned baffled": Muir, *John of the Mountains*, p. 416.

159 "No bears": *Ibid.*, p. 417.

159 "in its robes": *Ibid.*, pp. 417–18.

160 "a fitting conclusion": Merriam Diary.

160 "What's the matter": Muir, *John of the Mountains*, p. 418.

Chapter Twelve
"The Taking of the Totems"

162 "well furnished": Dellenbaugh Diary.

162 "all away getting gold": Muir, *John of the Mountains*, p. 419.

162 "Evidences of mining": Dellenbaugh Diary.

163 "The Vanished Tribe": Mrs. Edward Harriman, "Chronicles and Souvenirs."

164 "two of the half-grown": Muir, *John of the Mountains*, p. 419.

165 "with a view": Muir, *Travels in Alaska*, p. 74.

165 "The Taking of the Totems": Mrs. Edward Harriman, "Chronicles and Souvenirs."

168 "in fair preservation": Merriam Diary.

168 "love feast": Dall Diary.

170 "may have a meeting": Dellenbaugh Diary.

170 "he did not": *Ibid.*

170 "wild glee": Muir, *John of the Mountains*, p. 420.

Chapter Thirteen
"A Lapse of Time and a Word of Explanation"

171 "A Lapse of Time": Robert Service, *Ballads of a Bohemian* (Newark and New York: Barse and Hopkins, 1921).

171 had originally hoped: quoted in "The Harriman Expedition," New York *Daily Tribune*, August 15, 1889, p. 9.

171 "admirable management": "Discoveries in Our Arctic Region," *World's Work* I:149 (December 1900).

171 "an entire success": "The Harriman Expedition," New York *Daily Tribune*, August 15, 1899, p. 9.

172 the Blackfeet Reservation: Biographical information on George Bird Grinnell can be found in John F. Reiger, ed., *The Passing of the Great West: The Life and Letters of George Bird Grinnell* (New York: Winchester Press, 1972).

172 next sixteen years: Much of the information that follows on the careers of the scientists is obtained from *The Dictionary of American Biography* (New York: Charles Scribner's Sons, 1928–present) and *Biographical Memoirs*, 52 vols. (Washington, D.C.: National Academy of Science, 1928–1961).

173 "philistine idea": William Healey Dall to Frederick S. Dellenbaugh, November 1899, C. D. Wolcott Papers, Smithsonian Institution Archives.

173 "a large amount": Merriam Diary.

174 "Of all the great": Muir, *Edward Henry Harriman*, p. 3.

175 "I feel as if": Dellenbaugh Diary.

176 "four water-tight": *Ibid.*

176   "a plan for": Kennan, *E. H. Harriman's Far Eastern Plans*, p. 5.

177   "an all-land route": "Alaska's Railroads," *Review of Reviews* 29:346 (March 1904).

177   "in developing the country": Muir, *Edward Henry Harriman*, p. 27.

Chapter Fourteen
Edward S. Curtis' Alaskan Vision

181   taken as a hobby: Florence Curtis Graybill and Victor Boesen, *Edward Sheriff Curtis, Visions of a Vanishing Race* (N.Y.: Thomas Y. Crowell, 1976), p. 9; Curtis-Leitch Correspondence, Seattle Public Library.

183   Doré, before he shot it: Edward S. Curtis, "The Amateur Photographer," *The Western Trail* I, no. 5 (March–April 1900), p. 5.

184   luminist painters of the nineteenth century: see John Wilmerding, et al., *American Light: The Luminist Movement, 1850–1875, Paintings, Drawings, Photographs* (Washington, D.C.: National Gallery of Art, 1980).

192   hard taskmaster: C. Hart Merriam to Edward S. Curtis, letters, March to July 25, 1900, Merriam Letterbooks vol. I (1900), ms. in Bancroft Library.

192   studio in Seattle: *See* Curtis Studio advertisement, Photograph Collection, University of Washington.

Chapter Fifteen
Scientific Results

194   of the series: Harriman, *Harriman Alaska Series* vol. II, pp. 185–204.

194   "The geologic field": *Ibid.*, vol. IV, p. 6.

194   "the growth of": *Ibid.*, vol. III, p. 2.

195   "the restless motion": *Ibid.*, vol. I, p. 123.

195   "glorious views": *Ibid.*, p. 131.

195   volcanic archipelagoes: *Ibid.*, p. 129.

195   "the heavens opened": *Ibid.*, p. 130.

195   "The sail down": *Ibid.*, p. 135.

196   at the subject: *Ibid.*, pp. 170, 152.

196   "The rush to": *Ibid.*, p. 183.

196 "For hundreds of": *Ibid.*, vol. II, p. 138.

196 Lewis Henry Morgan: *See* Lewis Henry Morgan, *Ancient Society* (Cleveland: World Book Co., 1877) for the classic application of Herbert Spencer's Social Darwinism philosophy to anthropology.

196 "there is an inevitable": *Harriman Alaska Series* vol. I, p. 183.

197 "At present": *Ibid.*. II, p. 254.

197 they were abundant: *Ibid.*, p. 274.

197 "For the one Yosemite": *Ibid.*, p. 277.

197 "The Alaska coast": *Ibid.*

198 "the cash value": Expression used throughout William James, *Pragmatism: A New Name for Some Old Ways of Thinking* (New York: Longman's, Green and Co., 1907). The financial language was characteristic of the era. *See* Grove Karl Gilbert, "The Origin of Hypotheses," *Science*, n.s., III, no. 53 (1896), pp. 10–13, quoted in Stephen Pyne, "Grove Karl Gilbert: A Biography of American Geology," unpublished Ph.D. dissertation, American Civilization Program, University of Texas at Austin, 1976, p. 198, "Knowledge of Nature is an account at the bank, where each dividend is added to the·principal and the interest is ever compounded."

198 "of slates and shales": *Harriman Alaska Series* vol. IV, p. 7.

199 "the dominant constitution": *Ibid.*

199 Eocene or Jurassic: *Ibid.*, pp. 129–30.

199 time in Alaska: *Ibid.*, p. 7.

199 "the region was": *Ibid.*

200 publications of the age: *Ibid.*, vol. XIV, Editor's Preface.

200 Fifty new genera: This is based on reported discoveries of new species as they are claimed in the Harriman *Series*.

200 habitat of mollusks: *Harriman Alaska Series* vol. XIII, pp. 1–17.

200 in every category: See especially Gustave Eisen's work on the Enchytraeidae, which includes cross-section illustrations from dissections of worm "brains" and nervous systems. *Ibid.*, vol. XII, pp. 1–166.

201 not idle speculation: See Addison Emery Verrill's general discussion of starfishes in *Ibid.*, vol. XIV, pp. 1–19.

201 classical experiments: *See* Stephen Pyne, "Grove Karl Gilbert," *passim*. Much of the general discussion of Gilbert's thought is based on this work and discussions with Dr. Pyne.

202 "When an observer": *Harriman Alaska Series* vol. III, p. 113.

202 deemed the best: *Ibid.*, p. 5.
202 followed him: *Ibid.*, p. 2.
202 focus of his work: Pyne, "Grove Karl Gilbert," espec. pp. 373–79.
202 "glacial climate": *Ibid.*, p. 373.
202 affected Alaska: *Harriman Alaska Series* vol. III, p. 111.
203 "the combination": *Ibid.*, p. 109.
204 forward more easily: *Ibid.*, pp. 203–207.
204 to some extent: *Ibid.*
204 retreated recently: *Ibid.*, pp. 40–41.
204 of their creation: *Ibid.*, pp. 114–19, 143.
204 resistance and friction: *Ibid.*, pp. 160–61, 203–206, 220–23.
205 then the Pleistocene: *Ibid.*, pp. 113–39.
205 the ocean floor: *Ibid.*, pp. 210–18.
206 come to an end: For a general discussion of the "Second Great Age of Discovery," see William H. Goetzmann, "Paradigm Lost" in Nathan Reingold, ed., *The Sciences in the American Context: New Perspectives* (Washington, D.C.: Smithsonian Institution Press, 1979), pp. 21–34.

INDEX

Afogniak Islands, 117
Agriculture, U.S. Department of, 9,
18
Alaska Commercial Company, 86,
108
Amherst College (Mass.), 11
Annette Island, 38-43
Antarctic Continent, 88
Arctic Circle, 11
Audobon, John James, 197
Audobon Society, 156
Averell, Elizabeth, 19, 73, 99, 118,
144
Averell, Mrs. William, 101
Averell, William, 64, 97, 101

Badlands country, 21-22
Ballads of a Bohemian (Service), 171n
Barnum, P. T., 3
Barry Glacier, 110, 111, 112, 185
bear, Kodiak, 49, 118-22
Becker, G. F., 205
Berg Inlet, 79
Bering Sea (Straits), 4, 8, 126-28, 130,
131, 132, 133, 140, 141, 148, 149,
150, 159, 162, 177, 178
Bering, Vitus, 194
Berkeley, California, 193
Biological Survey, U.S. Department
of Agriculture, 8, 10, 61
Bishop Museum (Honolulu), 172
Bitter Creek, Wyoming, 19
Blackfeet Indians, 114

Blackfeet Reservation, 172
Blue Lakes (Shoshone Falls, Idaho),
23
Blue Mountains, 24
Boas, Franz, 196
Bogoslof Islands, 132, 133, 199
Boise, Idaho, 24; Natatorium, 24
Brady Glacier, 96
Brady, Governor, 89, 90, 91, 92, 93
Brewer, William H., 11, 16, 31, 33,
36, 38, 44, 49, 81, 92, 105, 112-13,
129, 149, 153, 157-58, 198
Britain, 109
Brooklyn Bridge, 4, 15
Bureau of American Ethnology, 41,
42
Burroughs, John, 4, 11, 18, 21-22, 26,
28, 30, 31, 33, 37, 38, 40, 41, 42,
44-45, 47, 48, 52, 53, 61, 62-63, 65,
72-73, 77, 81, 88, 90, 93, 95, 99,
100, 103-104, 106-107, 109, 112-
13, 117, 130, 132, 137, 146, 151,
157, 174, 195
Burt, Horace G., 20

California Academy of Sciences, 13
California Academy of Sciences
Museum (San Francisco), 13
California Geological Surveys, 11
Call of the Wild, The (London), 57
"Call of the Wild, The" (Service),
86n, 105n
Canada, 63, 64

Canyon Voyage, A (Dellenbaugh), 175
Cape Fox Village, 38, 162, 163, 165, 167
Cape Nome, 143
Carlisle, Pennsylvania, 41
Carnegie, Andrew, 6, 176
Cascade Reservation, 28
Catlin, George, 196
Century Club (New York), 16, 134, 175
Cheyenne, Wyoming, 19
Chicago, Illinois, 18
Chinese workers, exploitation of, 7, 104, 106, 113
Christian morality, 41
Chuchchis, Siberian, 141, 191
Church, Frederick, 183
City of Topeka (tourist ship), 93
Civil War, 22, 104
Clarence Strait, 44
coal mining, 19-20
Coast and Geodetic Survey, U.S., 194
Coe, Wesley R., 38, 76, 90, 101
Cole, Leon J., 75, 76, 122, 126
Colorado River, 3, 12, 20, 202
Columbia, The (steamship), 176
Columbia Glacier, 108, 183
Congress, U.S., 9
Cook, Captain James, 108, 194, 206
Cook Inlet, 114, 115, 116, 126, 156
Cooper Union (New York), 12
Cope, E. D., 22
copper mining, 109, 115
Copper River, 106-107
Cornell University, 13-14; Forestry School, 14
Cosmos Club (Washington, D.C.), 9, 10, 175
Coues, Elliott, 13, 148, 174-75
Coville, Frederick V., 9, 19, 20, 25, 38, 66, 76, 108, 111, 119, 157
Curtis, Edward S., 12, 29, 38, 47, 51, 64, 71, 73-74, 76, 77, 78, 87, 88, 99, 100, 101, 106, 111, 114, 121, 133, 137, 141, 145, 148, 150, 155, 162, 168, 171-72, 181-92, 202, 206

Dall, Annette, 39
Dall, William Healey, 8, 9, 32-33, 37, 38, 39, 43, 53, 62, 65, 66, 73, 89, 98, 99, 118, 119, 131, 136, 140-41, 149, 158, 168, 172-73, 194, 197, 198
Dalton Glacier, 97-98
Darwinism, social, 42, 58, 151, 201
Dawes Indian Severalty Act (1887), 42
Dawson City, Alaska, 61, 62
Dead Horse Trail, 60-61, 62
Death Valley expedition, 9, 10
deer, 49
Dellenbaugh, Frederick S., 7, 12-13, 14, 16, 18, 19, 20, 23, 25, 33, 37, 38, 43, 46, 50, 52-53, 59, 60, 62, 72, 78, 81, 83, 87, 91, 92, 96, 98, 101, 106, 107, 109, 112, 118, 119, 122, 123-25, 131, 134, 141, 144, 150, 151, 155, 156-57, 158, 159, 161-63, 168, 169, 173, 175
Devereux, Walter, 19, 50-51, 52, 97, 102, 115, 140, 144
Devil's Thumb, the, 48
Disenchantment Bay, 97-98
Doran, Captain Peter, 35, 79-80, 98, 105, 110, 133, 142, 151, 170
Doré, Paul, 183
Douglas Island, 52, 53, 162
Draper, Dorothea, 19, 73, 119, 123, 144
ducks, 48, 51
Duncan, William, 38-43
Dutch Harbor, 130-31, 151, 161
Dutton, Clarence, 199

eagles, 48-49, 119
Eagle Bay, 118
Egg Island, 98
Elliot, Daniel, 5, 38, 61, 111, 119, 144, 160
Emerson, Benjamin K., 11-12, 38, 44, 77, 198, 203
Emmons, George, 88, 89, 91
entomology. *See* Harriman Alaska Expedition: insect studies

Eskimos. *See* Indians
Explorer's Club, 175

Fairweather Range, 80, 81, 152, 159,
    195
Farragut Bay, 49, 50
Fernow, Bernhard E., 14, 19, 20, 23,
    25, 38, 44, 60, 66, 76, 80, 97, 123,
    147, 168, 197
Field Columbian Museum (Chicago),
    5, 61
Fish Commission, U.S., 194
Fisher, Albert K., 10, 13, 23, 37-38,
    43, 44, 51, 52, 76, 78, 80, 84, 95,
    133-34, 146, 148-50, 168
Foggy Bay, 163
Forest and Stream magazine, 156
Fortieth Parallel Survey, 10
"Fourth of July Ode" (Keeler), 123-24
fox, blue, 150
France, 109
Freud, Sigmund, 21
Fremont, John C., 22
Fuertes, Louis Agassiz, 7, 13, 18, 19,
    20-21, 37-38, 62, 76, 80, 83, 90-91,
    108, 122, 123, 126, 148-50, 157,
    174-75, 197

Gannett, Henry, 7, 9-10, 38, 60, 66,
    72, 73, 81, 83, 99, 102, 103, 104,
    109, 111, 118, 157, 197-98, 206
Geological Survey, U.S., 10, 176,
    194
George W. Elder (steamship), 4, 14, 29,
    30, 31-39, 43-44, 46-48, 50, 52-53,
    55-59, 67, 73, 74, 75, 79, 80, 83, 84,
    85, 89, 90, 93, 94, 95-103, 105, 107,
    108, 110, 114, 115, 116, 117, 121,
    122, 123, 126, 127, 129-45, 146-60,
    161-64, 168, 171, 184, 193, 194,
    206; as "floating hotel," 59; as
    "floating university," 6; grapho-
    phone aboard, 30, 92, 93, 112,
    123; nicknamed "George W. Roll-
    er," 33, 137
Gifford, Robert Swain, 12-13, 14, 16,

36, 37, 44, 72, 76, 90, 109, 112,
    119, 125, 144, 159, 161, 168, 175
Gilbert, Grove Karl, 8, 9, 23, 36, 38,
    70, 77, 78, 80, 81, 82, 84-85, 99,
    102, 103, 108, 111, 131, 172, 176,
    182, 184, 194-95, 198-99, 201-206
Glacier Bay, 59, 68, 69, 70, 72, 74,
    78, 117
Glaciers and Glaciation (Gilbert), 201-
    206
gold miners, 3, 43, 46, 56-61, 65-67,
    106-107, 130, 131, 143, 151, 163
gold rush, 7, 46, 55, 56, 62, 64, 65-67,
    143
Golofnin Bay, 108
Grand Canyon, 12
Grand Central Station (New York),
    16, 170
granite blasting, 61
graphophone. *See George W. Elder,*
    graphophone aboard
Great Plains, 21, 22
Greenland, 11
Green River, 20
Grinnell, George Bird, 12, 33, 38, 41,
    43, 49-50, 64-65, 69, 74, 89, 90, 91,
    97, 100, 102, 115, 135, 144-45, 156,
    160, 161, 172, 181-82, 195-96, 198
Gulf of Georgia (British Columbia),
    33, 44
gulls, 48, 148

Hall Island, 148
Harper's Publishing Company, 27
Harriman Alaska Expedition: animal
    studies, 49, 50; bear hunting, 5, 6,
    50, 73, 97, 102, 115, 116, 118-22,
    147; big-game hunting, 5, 69, 71,
    76, 90, 99, 117, 152, 158; bird and
    animal specimens, shooting and
    trapping of, 37-38, 45, 48, 62, 80,
    84, 97, 119, 130, 133-34, 146, 148,
    149, 164-65; bird studies, 37;
    choice of territory, 6; Committee
    on Literature and Art, 18; Com-
    mittee on Music and Entertain-
    ment, 123; Committee on Publica-

Harriman Alaska Expedition (*cont.*)
tion, 173; Executive Committee,
18, 32, 43; forestry studies, 66-67;
fossil collecting, 53; fur trading,
101, 108, 134-35, 137, 144, 155-56;
glaciers observed by, 12, 48, 81,
194, 201-206; insect studies, 83,
154; marine studies, 76, 154, 200-
201; mountains observed by, 43,
48, 58, 80, 152; photography, 51,
64, 71, 73-74, 99, 100, 106, 137,
150, 155, 181-92; press coverage of,
14, 28, 29-30, 137, 171; *Reports*,
193-200; salmon canning observed
by, 36, 106, 112-13, 155; seal
hunting, 78-79, 99-100, 188-89;
selection of members, 8-15;
sketching and painting, 25, 37, 38,
50, 83, 90-91, 106, 107, 109, 119,
151, 156; vegetation studies, 45, 66
Harriman, Averell, 4, 34, 47, 144,
165, 177-78
Harriman, Carol, 4, 38, 123, 124,
144, 154
Harriman, Cornelia, 4, 19, 38, 73, 81,
83, 99, 119, 123, 144, 146, 154,
155, 156
Harriman, Edward H., 3-15, 16-30,
32, 33, 34, 36, 38, 41, 47, 51, 53,
55, 56-58, 64, 65, 68, 69, 74, 80-81,
85, 89, 92, 108, 109, 110, 112, 114,
115, 118-28, 130, 132, 136, 140,
144-45, 146, 152, 162-63, 168-69,
172, 173-74, 176-77, 182, 191, 200,
206
Harriman Fjord, 111, 114
Harriman Glacier, 84, 112
Harriman, Mary, 4, 19, 38, 73, 81,
83, 90, 119, 144, 146, 154
Harriman, Mrs. E. H., 4, 38, 73,
119, 122, 127, 129, 144, 152, 156,
157, 200
Harriman, Roland, 4, 34, 47, 144,
155
Hayden, F. V., 9-10, 66
hot springs, 90
"Howling Valley" (Endicott Valley),
69, 71, 73, 76

Hubbard Glacier, 82, 97
Hudson River, 18, 45
Hugh Miller Glacier, 79, 80, 81, 84
hummingbirds, 37
Humphrey, Captain Omar J., 108,
110, 143

"Indian Jim," 97, 98, 108, 111, 157,
158
Indian River, 91
Indians (native Americans), 12, 37,
57, 79, 92, 182-83; Aleuts, 4;
Blackfeet, 114; canoes, 45, 112; as
cheap labor, 7, 41, 114; communities of, 36, 38-43, 46, 50-51,
54, 79-80, 84, 87-88, 163; Eskimos,
136-37, 140, 142, 143, 144, 158,
171, 189, 196; Kwakiutl, 196;
teaching of, 41-42; totem poles,
164-65, 168; Tlingits, 89, 97; and
whale hunting, 49; Yakutat, 125
industrial development, 41
Inverarity, D. J., 73-74, 81, 99, 111,
137, 182
Illinois Central Railroad, 5
Ithaca, New York, 193

Jackson, William H., 9
James, William, 198
jay bird, 45
Jordan, J. F., 31
Juneau, Alaska, 15, 48, 52, 53, 65, 67,
68, 160, 161

Kalama, Oregon, 29
Kearney, Thomas, 18, 38, 53, 75, 76,
84, 102, 104
Keeler, Charles, 13, 26-28, 29, 30, 31,
36, 38, 40, 44-45, 46, 47, 50, 81, 87,
92, 93, 103, 123-24, 126, 132, 134,
145, 146, 150, 151-52, 161, 172,
173, 174, 197
Keeler, Louise, 36
Kelly, Captain Luther "Yellowstone," 69, 97, 111, 119, 120-21,
129
Key to North American Birds (Coues),
13, 175

Kincaid, Trevor, 38, 53, 76, 83, 121,
129, 154, 156, 157, 163-64
King, Clarence, 10, 194
Klondike, the, 65, 131
Knowlton, F. H., 198
Kodak, 182
Kodiak. *See* St. Paul
Kodiak Island, 115, 116, 118, 120,
154, 198
Kandarkof, Stepan, 119, 121
Kukak Bay, 116
Kwakiutl Indians, 196

Lake Bennett, 62
La Perouse Glacier, 82, 96
LaTouche Island, 115
Lewiston, Idaho, 24, 25
Lind, Jenny, 3
London, Jack, 57-58
Lynn Canal, 55, 57, 68

Malaspina, Alejandro, 98
Malaspina Glacier, 96-100, 157, 195
Marsh, O. C., 22
Maybeck, Bernard, 174
Mazamas Club (Portland, Ore.), 28,
29
McClure, S. S., 27
McClure's Magazine, 27
Merriam, C. Hart, 8, 9, 10-14, 19, 20,
22, 23, 29, 36, 38, 44, 45, 50, 51,
52, 61, 69, 72, 75, 79, 89, 91, 97,
101, 102, 104, 115, 118, 119-21,
130, 133, 134-35, 144-45, 147, 149,
152, 154-55, 160, 168, 172, 173,
181-82, 192, 193, 198, 199-200,
202, 206
Metropolitan Club (New York), 9
Midsummer Night's Dream, A (Shakespeare), 154
Missouri Botanical Garden (St.
Louis), 10
Mohler, A. L., 24, 25, 29
Montana, 172
Moran, Thomas, 9
Morgan, J. Pierpont, 24, 29, 172, 176
Morgan, Lewis Henry, 196
Morris, Dr. Lewis Rutherford, 5, 38,

49, 50, 69, 97, 102, 112, 147, 157
mosquitoes, 84, 127
Mountains of California (Muir), 106
Mount Fairweather, 81, 83
Mount Rainier, 12, 181, 183
Mount St. Elias, 98, 158, 159, 195
Muir Glacier, 48, 59, 68, 69, 80, 81,
185, 191
Muir Inlet, 190
Muir, John, 4, 10-11, 19, 26-28,
29-30, 31, 32, 36, 38, 42, 44, 45,
46, 48, 49, 51, 56, 57, 59-60, 66, 68-
85, 90, 93, 99, 100-101, 102, 105,
106, 109, 110, 111, 119, 126, 130,
132, 134, 142, 144, 150, 152, 154,
159-60, 164, 170, 172, 173, 174,
177, 195, 206
Multnomah Falls, Idaho, 26
murres, 133-34

Nelson, Chaplain George F., 38, 39,
40, 144
New England, 21, 191
New Haven, Connecticut, 16
New Metlakahtla, 38-39, 53, 58
New York Daily Tribune, 171
New York Herald, 14
Nez Percé Indian Reservation
(Idaho), 24-25
Niagara Falls, 53
Nome, Alaska, 62, 131, 143
North American Commercial Company, 117, 130-31, 134, 155
North American Indian, The (Curtis),
181, 196
Northern Pacific Railroad, 23-25, 29
Northwest Passage, 98, 110, 112, 206
Nuchek, Alaska, 105
Nunatak, 186

Omaha, Nebraska, 20, 21
Orca, Alaska, 104, 106-108, 111, 113,
115, 143, 187
Oregon, 27
Oregonian (Portland, Ore.), 29, 171
Oregon Railroad and Navigation
Company, 14, 24
Oregon Trail, 24

ornithology. *See* Harriman Alaska Expedition: bird and animal specimens; bird studies
Osgood, Wilfred, 61, 62, 175
Otis Elevators, 15
"Our Banner" (Dellenbaugh), 124
owl, snowy, 150

Pabst Blue Ribbon beer, 23-24
Pacific Coast Steamship Company, 6
Pacific Coast Whaling Company, 113
Pacific Glacier, 78
Pacific Steam Whaling Company, 106, 108, 113, 143
Page, Walter Hines, 26
palace cars (private Pullman railroad cars), 17, 20, 170
Palache, Charles, 12, 19, 38, 44, 51, 53, 78, 80, 84, 85, 91, 97, 108, 111, 119, 129, 146, 153, 175-76, 198
Panama Canal, 15
Paris Exposition, 109
Patterson Glacier, 48
Pavlof Volcano, 130
Peril Strait, 86
Pinchot, Gifford, 172, 173, 181
Plover Bay, 135-37, 138-39, 141, 188, 196
Point Gustavus, 76-78, 80, 83, 84
Pollock, Jackson, 184
Popof Island, 129
Porcupine Hill, 60
Port Clarence, 140, 141, 142, 143, 144, 145, 146, 196
Porter, Bishop, 38
Portland Heights (Portland, Ore.), 28
Portland Hotel (Portland, Ore.), 28
Portland, Oregon, 26, 28, 172
Port Wells, 109, 112, 115
Powell expedition, 175
Powell, John Wesley, 3, 12, 20, 41, 194
Pribilof Islands, 134
Princess Royal Island, 36; Lowe Inlet, 36
Prince William Sound, 105, 106, 108, 109, 111, 117, 198
Protestant work ethic, 41

Puget Sound, 66, 181, 191, 197
Pyne, Stephen, 201

quartz mining, 51-53

railroad around the world, 4, 8, 127-28, 176-77, 193
Rainier Club (Seattle, Wash.), 29
Rainier Park, 28
ravens, 37-38
Reid Inlet, 70, 80
Ridgway, Robert, 10, 23, 38, 76, 103, 156
"Rime of the Ancient Mariner, The" (Coleridge), 103
Ritter, William E., 13, 38, 53, 76, 79, 83, 87, 90, 101, 103, 104, 129, 172
Rockefeller, Nelson, 178
Rocky Mountains, 19, 22, 202
Roosevelt, Theodore, 5, 11, 12, 173, 175
Rosalie (mail boat), 58
Ruskin, John, 23, 63, 195
Russia, 109
Russian American Company, 87

St. Lawrence Island, 146, 148
St. Matthews Island, 148-50
St. Paul Island, 134
St. Paul (Kodiak), 117, 118, 119, 122, 126
Saldovia Point, 115
Salt Lake City, Utah, 12-13
sandpipers, 148
Sand Point, 153, 154
San Francisco, California, 13, 26, 27, 28, 31, 57, 93, 106, 107, 113, 131, 146, 151, 173; Bay Area, 12, 172
Saunders, Alton, 38, 45, 53, 76, 83, 90, 103, 129, 154
Seattle, Washington, 3-4, 14, 29, 74, 146, 153, 171, 172, 206
seismology, 12
Senate, U.S., 55
Service, Robert, 86n, 105n, 116n, 171n
Seymour's Narrows, 33
"Shadow-Catcher, The" (Curtis), 192

"Shooting of Dan McGrew, The" (Service), 116*n*
Shoshone Falls, Idaho, 22-23
Shumagin Islands, 117, 127, 129
Siberia, 8, 126-27, 131, 132, 136, 140, 143
Simple Home, The (Keeler), 174
Sitka, Alaska, 86-94, 95, 97, 119, 132
Skagway, Alaska, 15, 53, 55-67, 87
Skagway River, 60
Smith, "Soapy," 57
Smithsonian Institution (Washington, D.C.), 10, 146, 147, 172, 173, 200
Snake River, 23, 24-26
Snake River Canyon, 23
Souvenir Album (Curtis), 191-92
Spanish-American War, 7, 123
sparrows, 45; golden-crowned, 62, 130
Spell of the Yukon, The (Service), 86*n*, 105*n*, 116*n*
Stanley-Brown, J., 131, 157, 165
Starks, Edwin C., 83
"Stars and Stripes Forever" (Sousa), 123
Steel (President, Mazamas Club), 28
Stickeen River, 46
Stikeen (dog), 59, 71
Suez Canal, 15
swallows, 45
swans, 147

"Taking of the Totems, The" (Stanley-Brown), 165
Taku Village, 50-51
Tentlatch, Chief, 89, 91, 92
Thomas, Lowell, 175
Thumb, Tom, 3
Tlingit Indians, 89
totem poles, 164, 168
tourism, 6, 52, 55, 65, 66, 71-72, 86, 92
Trans-Siberian Railroad, 127-28
Treadwell Mine (Juneau), 50-53, 54, 56, 198
Trelease, William, 10, 38, 119

Trudeau, Edward, 19, 69, 76, 97, 101, 112, 150

Ulrich, E. O., 198, 199
Unalaska Island, 130, 131, 132, 137, 150, 151
Union Pacific Railroad, 5, 6-7, 20, 24, 114
United States, 63-64, 109, 123, 171
United States Exploring Expedition (1838-1842), 88
University of California at Berkeley, 13
University of Michigan, 168
University of Washington, 154
"Utopia" (private railroad car), 17, 26
Uyak Bay, 117

Vancouver Island, 32-33
"Vanished Tribe, The" (Kincaid), 163
Verill, Addison Emery, 200-201
Victoria, British Columbia, 32
Victoria Museum (Victoria, B.C.), 32
Von Richtofen (expert on volcanoes), 199

warblers, 37
Washburn, M. L., 108, 117, 155-56, 198
Washington Academy of Sciences (D.C.), 9, 14
Washington, D.C., 43, 68, 116, 150, 193
Washington Indians, 172
whales, 49, 143
White Pass City, 62
White Pass Railroad, 56-57, 60, 64, 128
White Pass Ridge, 61, 63, 158, 175
Whitman, Walt, 11, 21, 107
Willamette River, 29
Wind River Mountains, 20
Wood Island, 122-23
Wordsworth, William, 134
World's Columbian Exposition (Chicago), 15
World's Work (newspaper), 171

Wrangell, Alaska, 45, 46, 47, 54, 113, 163
Wrangell Narrows, 47
Wyoming, 21

Yakutat Bay, 91, 95, 96, 97, 100, 101, 103, 104, 117, 158-59, 188, 198
Yakutat Indians, 125
Yakutat Village, 95, 96, 97

Yale University, 11
Yellowstone National Park, 9, 66, 197
Yosemite National Park, 173, 197
Young, Hall, 59-60, 71, 85
Yukon River, 57, 62
Yukon territory, 8, 57, 61, 96, 143

zoology. *See* Harriman Alaska Expedition: animal studies